Miller's Guide to Home Wiring

Home Reference Series

Miller's Guide to Home Wiring

REX MILLER
Professor Emeritus
State University College at Buffalo
Buffalo, New York

MARK R. MILLER
Associate Professor
Texas A&M University–Kingsville
Kingsville, TX

GLENN E. BAKER
Professor Emeritus
Texas A&M University
College Station, TX

McGraw-Hill

New York   Chicago   San Francisco   Lisbon   London
Madrid   Mexico City   Milan   New Delhi   San Juan
Seoul   Singapore   Sydney   Toronto

The McGraw·Hill Companies

Library of Congress Cataloging-in-Publication Data

Miller, Rex.
    Wiring the house / Rex Miller.
      p. cm.
    ISBN 0-07-144551-X
    1. Electric wiring, Interior.   2. Dwellings—Electric equipment.   I. Title.

    TK3285.M545   2004
    621.329'24—dc22

                                             2004054870

1 2 3 4 5 6 7 8 9 0   QPD/QPD   0 1 0 9 8 7 6 5 4

ISBN 0-07-144551-X

The sponsoring editor for this book was Larry Hager, the editing supervisor was Caroline Levine, and the production supervisor was Sherri Souffrance. It was set in ITC Century Light by Wayne Palmer of McGraw-Hill Professional's Hightstown, N.J., composition unit.

McGraw-Hill books are available at special quantity discounts to use as premiums and sales promotions, or for use in corporate training programs. For more information, please write to the Director of Special Sales, McGraw-Hill Professional, Two Penn Plaza, New York, NY 10121. Or contact your local bookstore.

Contents

12 Appliances, Ovens, Ranges, Cooking Tops, Air Conditioners, Cable TV, and Internet

Appendix. Tables

Preface

The purpose of this book is to aid those who are interested in the subject of wiring a house, remodeling an existing home, and/or becoming an electrician.

This book has something for the apprentice, the seasoned electrician, and the do-it-yourselfer. Although it contains some theory of the principles of electricity, the main emphasis of the text is on the practical, everyday applications of those principles.

Many illustrations are included to show the variety of parts and techniques presently used as well as parts found in older homes that may need replacement or repair. Obviously, not all related problems will be presented here since there is a great deal of ingenuity required of the worker on the job. For standard procedure, however, the National Electrical Code *Handbook* does give guidance to the number of wires in a box, the types of wire and where to use them, and the kinds of equipment that are safe to install in a given type of location. The *Handbook* should be part of your toolbox.

Keep in mind that the *Code,* as it is commonly called, changes from time to time. In fact, every three years a new *Code* is published with changes to the old. It is in a constant state of flux and is changed as conditions warrant. Therefore, it may be possible that information related to various box capacities and equipment will change with the times. It is imperative that an electrician be aware of the *Code* and its application to the local situation.

REX MILLER
MARK R. MILLER
GLENN E. BAKER

Acknowledgments

AS IS THE CASE WITH ANY BOOK, MANY people have given generously of their time in assisting the author. Their efforts and suggestions have been of great value. I would like to take this opportunity to express my appreciation.

Many manufacturers of electrical parts and equipment have contributed illustrations, written materials, and suggestions for inclusion in this book. They have done so with a great deal of professionalism.

Certain manufacturers have been identified with the field for many decades and have some very good illustrations of their product lines. The following list gives their names. These manufacturers and agencies are but a sampling of the many who make the generation, installation, and use of electricity as efficient and safe as possible. The technical data furnished by these companies and agencies are much appreciated. Without their assistance, this project would have had neither appeal or objectivity.

Appleton Electric Company
Bernzomatic Corporation
Brown & Sharpe Manufacturing Company
Bryant Electric Company
Buchanan Electric Products Division
Canadian Standard Association, CSA
Central Illinois Light and Power, CILCO
Circle F Industries
Edison Electric Institute, NYC
Eller Manufacturing Co.
Emerson Electric Co.
General Electrical Company
Greenlee Tool Company
Harvey Hubbell, Inc.

Heath Company
Honeywell
Ideal Industries, Inc.
Misener Manufacturing Co.
National Fire Protection Association
National Safety Council
New York Board of Fire Underwriters
New York State Power Authority
Niagara Mohawk Power Corporation, NMP
NuTone Division of Scovill
Ohmite
Onan
Pass and Seymour, Inc.
Rural Electrification Authority, REA
Sangamo Electric Company
Sears, Roebuck and Company
Seatek Company
Slater Electric Co.
Square D. Company
Superior Electric Company
Thomas & Betts Company
3M, Electro-Products Division
Underwriters' Laboratories, Inc.
Union Insulating Company
U.S. Department of Labor
Weston Electrical Instruments Co.

The following individuals also helped with the preparation of this book:

Jim Kasprzyk, illustrator, did the drawings.

Ed Zempel aided in the editorial work.

1
CHAPTER

Electricity and Electrical Circuits

WHAT IS ELECTRICITY?

Although you cannot see electricity, you are aware of it every day. You see it used in countless ways. You cannot taste or smell electricity, but you can feel it. You can *taste* food cooked with its energy. You can *smell* the gas (ozone) that forms when lightning passes through the air.

Basically there are two kinds of electricity—*static* (stationary) and *current* (moving). This book is chiefly about current electricity. That is the kind of electricity commonly put to use.

Current electricity can be simply defined as the *flow of electrons along a conductor.* To understand that definition, you must know something about chemical elements and atoms.

METRIC MEASUREMENT IN ELECTRICITY

Most of the units used in electricity (such as ampere and volt) are metric units. However, linear and volume measurements in the National Electrical Code are given in U.S. Customary System (USCS) measurement. When the SI metric system is wholly adopted in the United States, the linear and volume measurements used in the Code will probably be stated in metric units. However, such a change is perhaps several years away. For this reason, measurements of length and volume in this text have been given in USCS units. Conversions to metric have not been made. Table 1-4 (page 22) gives the information necessary to convert USCS measurements to metric measurements, and vice versa.

In the metric system of measurement, in numbers of five digits and more, a space (rather than a comma) is inserted every three digits, counting from the decimal point. Thus, 10,000 is written 10 000. The same rule applies to decimals: 0.000001 is written as 0.000 001.

ELEMENTS AND ATOMS

Elements are the most basic materials in the universe. Ninety-four elements, such as iron, copper, and nitrogen, have been found in nature. Scientists have made 11 others in laboratories. Every known substance—solid, liquid, or gas—is composed of elements.

It is very rare for an element to exist in a pure state. Nearly always the elements are found in combinations called *compounds.* Even such a common substance as water is a compound, rather than an element (Fig. 1-1).

An *atom* is the smallest particle of an element that retains all the properties of that element. Each element has its own kind of atom. Thus, all hydrogen atoms are

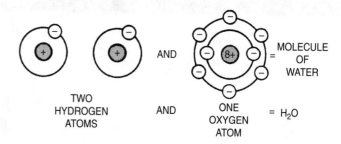

Fig. 1-1 *Two or more atoms linked are called a molecule. Here are two hydrogen atoms and one oxygen atom that form a molecule of the compound water (H_2O).*

alike. They are different from the atoms of all other elements. However, all atoms have certain things in common. They all have an inner part, the *nucleus.* This is composed of tiny particles called *protons* and *neutrons.* An atom also has an outer part. It consists of other tiny particles, called *electrons.* The electrons orbit the nucleus (Figs. 1-2 and 1-3).

Neutrons have no electrical charge, but protons are positively charged. Electrons have a negative charge. Because of these charges, protons and electrons are particles of energy. That is, these charges form an electric field of force within the atom. Stated very simply, these charges are always pulling and pushing each other. This makes energy in the form of movement.

The atoms of each element have a definite number of electrons. They have the same number of protons. For example, a hydrogen atom has one electron and one proton. An aluminum atom has 13 of each. The opposite charges—negative electrons and positive protons—attract one another and tend to hold electrons in orbit. As long as this arrangement is not changed, an atom is electrically balanced.

However, the electrons of some atoms are easily pushed or pulled out of their orbits. This ability of electrons to move or flow is the basis of current electricity.

Free Electrons

In some materials, heat causes electrons to be forced loose from their atoms. In other materials, such as copper, electrons may be easily forced to drift, even at room temperatures. When electrons leave their orbits, they may move from atom to atom at random, drifting in no particular direction. Electrons that move in such a way are referred to as *free electrons.* However, a force can be applied to direct them in a definite path.

- *Insulators.* An insulator is a substance that restricts the flow of electrons. Such materials have a very limited number of free electrons. Thus, you see that the movement of free electrons classifies a material as

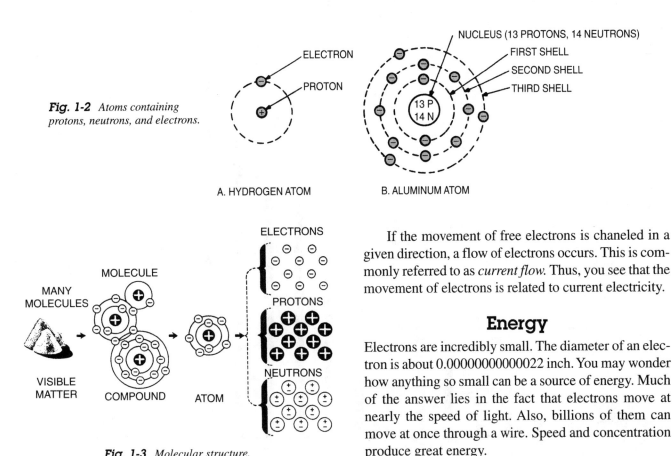

Fig. 1-2 *Atoms containing protons, neutrons, and electrons.*

A. HYDROGEN ATOM

B. ALUMINUM ATOM

Fig. 1-3 *Molecular structure.*

either a conductor or an insulator. No material is known to be a perfect insulator, that is, entirely void of free electrons. However, there are materials which are such poor conductors that for all practical purposes they are placed in the insulator class.

Wood, glass, mica, and polystyrene are insulators (Fig. 1-4). They have varying degrees of resistance to the movement of their electrons. The higher the line on the chart in Fig. 1-4, the better are the insulating qualities of the material.

• *Semiconductors.* You have heard the word *semiconductor* used in relation to transistors and diodes in electronic equipment. Materials used in the manufacture of transistors and diodes have a conductivity between that of a good conductor and that of a good insulator. Therefore, the name *semiconductor* is given to them. Germanium and silicon are the two most commonly known semiconductor materials. Through the introduction of small amounts of other elements, these nearly pure (99.999999%) elements become limited conductors. The manufacture of semiconductors is a fascinating process. However, it is too detailed to discuss here. You may wish to research the topic on your own by checking out a book from your library.

If the movement of free electrons is chaneled in a given direction, a flow of electrons occurs. This is commonly referred to as *current flow.* Thus, you see that the movement of electrons is related to current electricity.

Energy

Electrons are incredibly small. The diameter of an electron is about 0.00000000000022 inch. You may wonder how anything so small can be a source of energy. Much of the answer lies in the fact that electrons move at nearly the speed of light. Also, billions of them can move at once through a wire. Speed and concentration produce great energy.

ELECTRICAL MATERIALS
Conductors

A conductor is a material through which electrons move. Actually, all metals and most other materials are conductors to some extent. Some, however, are better than others. Thus, the term *conductor* is usually used to mean a material through which electrons move freely.

What makes one material a better conductor than another? A material that has many free electrons tends to be a good conductor. For practical purposes, however, there are other points that must be considered in choosing a material to use as a conductor.

For example, gold, silver, aluminum, and copper are good conductors. However, the cost of gold and silver limits their use. Copper, because of its superior strength in both hot and cold weather, is preferred over aluminum for many uses.

GENERATING ELECTRICITY

There are several ways to produce electricity. Remember: *Electricity is the flow of electrons along a conductor.* Friction, pressure, heat, light, chemical action, and magnetism are among the more practical methods used to make electrons move along a conductor. Other methods

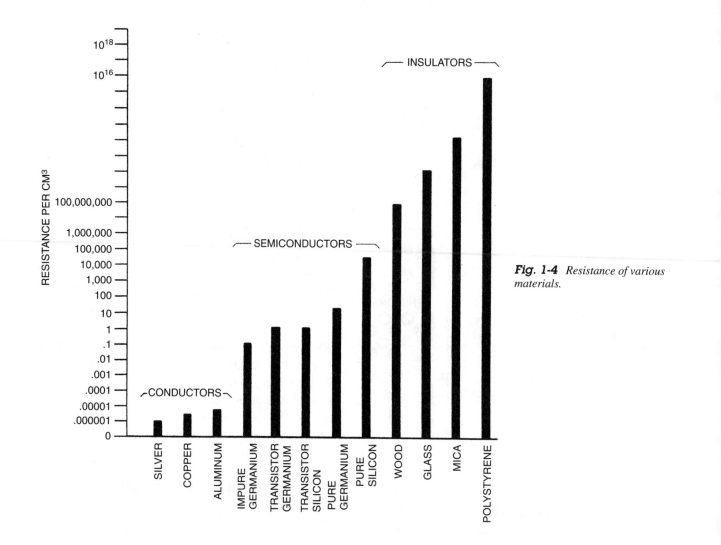

Fig. 1-4 *Resistance of various materials.*

(sometimes called *exotic*) are used to generate electricity for special purposes. For instance, experimental cells developed for the space program are termed *exotic*.

- *Friction.* Electricity is produced when two materials are rubbed together. The movement of your shoes against the carpet can cause static electricity. Some practical applications of static electricity are in the manufacture of sandpaper and in the cleaning of polluted air (Fig. 1-5).

- *Pressure.* Electricity is produced when pressure is applied to a crystal. The crystals are usually Rochelle salts or quartz. This is also known as the *piezoelectrical effect.*

- *Heat.* Electricity is produced when heat is applied to the junction of two dissimilar metals. This junction is usually referred to as a *thermocouple.* The thermocouple is used to measure temperatures in industrial applications. This is especially true in checking the temperature of kilns for ceramic work (Fig. 1-6).

- *Light.* Electricity is produced when light strikes a photosensitive material. (The word *photo* means

light.) Photoelectric cells are used in cameras, spacecraft, and in radios (Fig. 1-7).

- *Chemical action.* Electricity is produced when a chemical action takes place between two metals in a cell. A single unit is called a cell. Connecting two or more cells together produces a battery. Batteries are used in flashlights, radios, hearing aids, and calculators. The automobile uses a lead-acid cell combination. You could not start today's cars without a battery. Many types of cells are available today (Fig. 1-8).

- *Magnetism.* Electricity is produced when a magnet is moved past a piece of wire. Or, a piece of wire can be moved through a magnetic field. The result is the same. Motion, a magnetic field, and a piece of wire are needed to produce electricity. To date, magnetism is the most inexpensive way of producing electrical power. We use magnetism to produce electricity for homes.

- *Exotic generators.* The fuel cell is one of the latest developments for the production of electricity. The

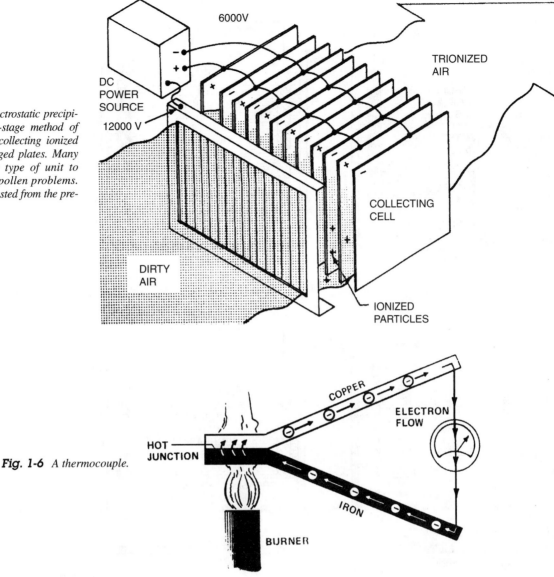

Fig. 1-5 *The electrostatic precipitator uses a two-stage method of cleansing air by collecting ionized particles on charged plates. Many homes have this type of unit to reduce dust and pollen problems. Clean air is exhausted from the precipitator.*

6000V

TRIONIZED AIR

DC POWER SOURCE

12000 V

COLLECTING CELL

DIRTY AIR

IONIZED PARTICLES

Fig. 1-6 *A thermocouple.*

COPPER

ELECTRON FLOW

HOT JUNCTION

IRON

BURNER

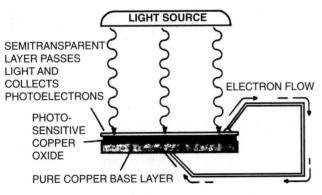

LIGHT SOURCE

SEMITRANSPARENT LAYER PASSES LIGHT AND COLLECTS PHOTOELECTRONS

ELECTRON FLOW

PHOTO-SENSITIVE COPPER OXIDE

PURE COPPER BASE LAYER

Fig. 1-7 *A photoelectric cell.*

oxygen-concentration cell includes an electrolyte. The electrolyte conducts an electric charge in the form of oxygen ions, but acts as an insulator to electrons. The electrolyte is located between two electrodes. (The electrolyte is wet. Electrodes are usually metal rods or sheets.) By causing oxygen of different concentrations to pass by the electrodes, it is possible to produce electricity.

Current is assumed to flow from negative (−) to positive (+) terminals of a battery or generator.

Current is measured in amperes. In electronics, it is sometimes necessary to use smaller units of measurement. The *milliampere* (abbreviated as mA) is used to indicate one one-thousandth of an ampere (0.001 A). If an even smaller unit is needed, it is usually the *microampere* (μA). The microampere is one-millionth of an ampere. This may be written as 0.000001 A. The Greek letter *mu* (μ) is used to indicate *micro*. (Table 1-1 lists the letters of the Greek alphabet and the terms they designate.)

A voltmeter is used to measure voltage. An ammeter is used to measure current in amperes. A microammeter

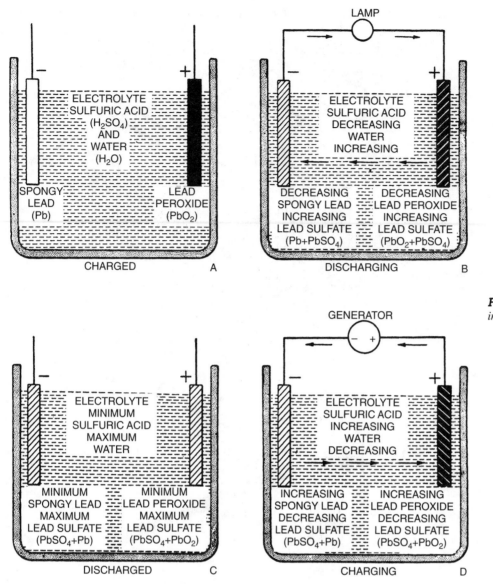

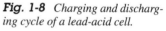

Fig. 1-8 *Charging and discharging cycle of a lead-acid cell.*

or a milliammeter may be used to measure smaller units of current.

Resistance

The movement of electrons along a conductor meets with some opposition. This opposition is *resistance.* Resistance is useful in electrical and electronics work. Resistance makes it possible to generate heat, control electron flow, and supply the correct voltage to a device.

Resistance in a conductor depends on four factors: material, length, cross-sectional area, and temperature.

- *Material.* Some materials offer greater resistance than others. It depends upon the number of free electrons present in the material.

- *Length.* The longer the wire or conductor, the greater resistance it has. The resistance is said to vary directly with the length of the wire.

- *Cross-sectional area.* Resistance varies inversely with the size of the conductor in cross section. In other words, the larger the wire, the smaller the resistance per foot of length.

- *Temperature.* For most materials, the higher the temperature, the higher the resistance. However, there are some exceptions to this in devices known as *thermistors.* Thermistors change resistance with temperature. They decrease in resistance as the temperature increases. Thermistors are used in meters and as temperature indicators.

Resistance is measured by a unit called the *ohm.* The Greek letter *omega* (Ω) is used as the symbol for

Table 1-1 *The Greek Alphabet*

Name	Capital	Small	Used to Designate
Alpha	A	α	Angles, area, coefficients, and attenuation constant
Beta	B	β	Angles and coefficients
Gamma	Γ	γ	Electrical conductivity and propagation constant
Delta	Δ	δ	Angles, increment, decrement, and determinants
Epsilon	E	ε	Dielectric constant, permittivity, and base of natural logarithms
Zeta	Z	ζ	Coordinates
Eta	H	η	Efficiency, hysteresis, and coordinates
Theta	Θ	ϑ, θ	Angles and angular phase displacement
Iota	I	ι	Coupling coefficient
Kappa	K	κ	
Lambda	Λ	λ	Wavelength
Mu	M	μ	Permeability, amplification factor, and prefix *micro*
Nu	N	ν	
Xi	Ξ	ξ	
Omicron	O	o	
Pi	Π	π	Pi = 3.1416
Rho	P	ρ	Resistivity and volume charge density
Sigma	Σ	σ, ς	Summation
Tau	T	τ	Time constant and time-phase displacement
Upsilon	Υ	υ	
Phi	Φ	ϕ, φ	Magnetic flux and angles
Chi	X	χ	Angles
Psi	Ψ	ψ	Dielectric flux
Omega	Ω	ω	Resistance in ohms and angular velocity

electrical resistance. The letters of the Greek alphabet and their designations are shown in Table 1-1.

The hydrogen-oxygen cell produces water and electricity. Such a cell was used on one of the spaceflights to supply drinking water and electricity in a very small space. Other exotic cells—not all of them perfected yet—are the redox fuel cell, the hydrocarbon fuel cell, the ion-exchange membrane, and the magnetohydrodynamic (MHD) generator (Fig. 1-9). In the MHD generator, hot plasma is generated and seeded in a burner similar to a rocket engine. It then travels through a magnetic field applied at right angles to the flow and past electrodes exposed to this stream of gas. Electrons in the gas are deflected by the field. Between collisions with the particles in the gas, they make their way to one of the electrodes. Electricity flows as the electrons move from the cathode, through the load, to the anode, and back again to the gas stream. There are thousands of other methods of producing electricity. Some of them will now find development funds because of the energy crisis.

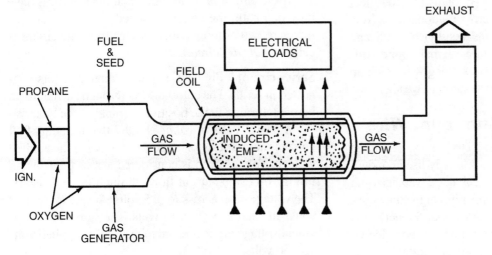

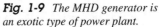

Fig. 1-9 *The MHD generator is an exotic type of power plant.*

VOLTAGE AND CURRENT

So far you have become aware of what electricity is. You have learned some of the ways it is produced. Now it is time to learn how electrical energy is measured. The units of measurement most frequently used are *voltage* and *current*.

- *Voltage.* We measure the difference in potential between two plates in a battery in terms of volts. It is actually *electrical pressure* exerted on electrons in a circuit. (A circuit is a pathway for the movement of electrons.) An external force exerted on electrons to make them flow through a conductor is known as *electromotive force,* or emf. It is measured in volts. Electrical pressure, potential difference, and emf mean the same thing. The terms *voltage drop* and *potential drop* can be interchanged.

- *Current.* For electrons to move in a particular direction, it is necessary for a potential difference to exist between two points of the emf source. If 6,280,000,000,000,000,000 electrons pass a given point in 1 second, there is said to be 1 *ampere* of current flowing. The abbreviation for ampere is A. The same number of electrons stored on an object (static charge) and *not moving* is called a *coulomb* (abbreviated as C).

Wire Size

As you become more familiar with electricity and its circuits and with some of the requirements for wiring a house or building, you will become more aware of the current-carrying abilities of wire. The size of the wire is given in numbers. This size usually ranges from 0000 (referred to as *four-ought*) to No. 40. *The larger the wire, the smaller its number.*

For instance, No. 32 wire is smaller than No. 14. Table 1-2 shows the resistance [in ohms per 1000 feet (305 meters)] in relation to the cross-sectional area. Note how the temperature affects the resistance at 25 and 65°C (77 and 149°F). Temperature can make quite a difference in resistance for long wires. Long wires pick up heat and expand when exposed to summer weather.

Copper versus Aluminum Wire

Although silver is the best conductor, its use is limited because of its high cost. Two commonly used conductors are aluminum and copper. Each has advantages and disadvantages. For instance, copper has high conductivity and is more ductile (can be drawn out thinner). It is relatively high in tensile strength and can be soldered easily. But it is more expensive than aluminum.

Aluminum has only about 60% of the conductivity of copper. It is used in high-voltage transmission lines and sometimes in home wiring. Its use increased in recent years. However, most electricians will *not* use it to wire a house today. There are several reasons for this. These will become apparent as we progress through the book.

If copper and aluminum are twisted together, as in a wire nut connection, it is possible for moisture to get to the open metals. Corrosion will take place, causing a high-resistance joint. This can result in a dimmer light or a malfunctioning motor.

Circuits

Complete circuit   A complete circuit is necessary for the controlled flow or movement of electrons along a conductor (Fig. 1-10). A complete circuit is made up of a *source of electricity,* a *conductor,* and a *consuming device.* The flow of electrons through the consuming device produces heat, light, or work.

To form a complete circuit, these rules must be followed (refer to Fig. 1-14):

1. Connect one side of the power source to one side of the consuming device (A to B).
2. Connect the other side of the power source to one side of the control device, usually a switch (C to D).
3. Connect the other side of the control device to the consuming device that it is supposed to control (E to F).

This method is used to make a complete path for electrons to flow from one terminal of the battery or power source containing an excess of electrons to the terminal that has a deficiency of electrons. The movement of the electrons along the completed path provides energy. Of course, in order for the path to be complete, the switch must be closed. (Refer to Fig. 1-10.)

If the circuit is so arranged that the electrons have only one path, the circuit is called a *series circuit.* If there are two or more paths for electrons, the circuit is called a *parallel circuit.*

Series circuit   Figure 1-11 shows three resistors connected in series. The current flows through each of them before returning to the positive terminal of the battery.

Kirchoff's law of voltages states that the sum of all voltages across resistors or loads is equal to the applied voltage. Voltage drop is considered across the resistor. In Fig. 1-11, the current flows through three resistors. The voltage drop across R_1 is 5 volts. Across R_2, it is 10 volts, and across R_3 it is 15 volts. The sum of the individual voltage drops is equal to the total or applied voltage, 30 volts.

Table 1-2 *Standard Annealed Solid Copper Wire (American Wire Gage—B & S)*

| Gage number | Diameter (mils) | Cross Section | | Ohms per 1000 feet | | Ohms per Mile at 25°C (77°F) | Pounds per 1000 feet |
		Circular mils	Square inches	25°C (77°F)	65°C (149°F)		
0000	460.0	212,000.0	0.166	0.0500	0.0577	0.264	641.0
000	410.0	168,000	0.132	0.0630	0.0727	0.333	508.0
00	365.0	133,000.0	0.105	0.0795	0.0917	0.420	403.0
0	325.0	106,000.0	0.0829	0.100	0.116	0.528	319.0
1	289.0	83,700.0	0.0657	0.126	0.146	0.665	253.0
2	258.0	66,400.0	0.0521	0.159	0.184	0.839	201.0
3	229.0	52,600.0	0.0413	0.201	0.232	1.061	159.0
4	204.0	41,700.0	0.0328	0.253	0.292	1.335	126.0
5	182.0	33,100.0	0.0260	0.319	0.369	1.685	100.0
6	162.0	26,300.0	0.0206	0.403	0.465	2.13	79.5
7	144.0	20,800.0	0.0164	0.508	0.586	2.68	63.0
8	128.0	16,500.0	0.0130	0.641	0.739	3.38	50.0
9	114.0	13,100.0	0.0103	0.808	0.932	4.27	39.6
10	102.0	10,400.0	0.00815	1.02	1.18	5.38	31.4
11	91.0	8,230.0	0.00647	1.28	1.48	6.75	24.9
12	81.0	6,530.0	0.00513	1.62	1.87	8.55	19.8
13	72.0	5,180.0	0.00407	2.04	2.36	10.77	15.7
14	64.0	4,110.0	0.00323	2.58	2.97	13.62	12.4
15	57.0	3,260.0	0.00256	3.25	3.75	17.16	9.86
16	51.0	2,580.0	0.00203	4.09	4.73	21.6	7.82
17	45.0	2,050.0	0.00161	5.16	5.96	27.2	6.20
18	40.0	1,620.0	0.00128	6.51	7.51	34.4	4.92
19	36.0	1,290.0	0.00101	8.21	9.48	43.3	3.90
20	32.0	1,020.0	0.000802	10.4	11.9	54.9	3.09
21	28.5	810.0	0.000636	13.1	15.1	69.1	2.45
22	25.3	642.0	0.000505	16.5	19.0	87.1	1.94
23	22.6	509.0	0.000400	20.8	24.0	109.8	1.54
24	20.1	404.0	0.000317	26.2	30.2	138.3	1.22
25	17.9	320.0	0.000252	33.0	38.1	174.1	0.970
26	15.9	254.0	0.000200	41.6	48.0	220.0	0.769
27	14.2	202.0	0.000158	52.5	60.6	277.0	0.610
28	12.6	160.0	0.000126	66.2	76.4	350.0	0.484
29	11.3	127.0	0.0000995	83.4	96.3	440.0	0.384
30	10.0	101.0	0.0000789	105.0	121.0	554.0	0.304
31	8.9	79.7	0.0000626	133.0	153.0	702.0	0.241
32	8.0	63.2	0.0000496	167.0	193.0	882.0	0.191
33	7.1	50.1	0.0000394	211.0	243.0	1,114.0	0.152
34	6.3	39.8	0.0000312	266.0	307.0	1,404.0	0.120
35	5.6	31.5	0.0000248	335.0	387.0	1,769.0	0.0954
36	5.0	25.0	0.0000196	423.0	488.0	2,230.0	0.0757
37	4.5	19.8	0.0000156	533.0	616.0	2,810.0	0.0600
38	4.0	15.7	0.0000123	673.0	776.0	3,550.0	0.0476
39	3.5	12.5	0.0000098	848.0	979.0	4,480.0	0.0377
40	3.1	9.9	0.0000078	1,070.0	1,230.0	5,650.0	0.0299

To find the total resistance in a series circuit, just add the individual resistances ($R_T = R_1 + R_2 + R_3 + \cdots$).

Parallel circuit In a parallel circuit, each load (resistance) is connected directly across the voltage source. There are as many separate paths for current flow as there are branches (Fig. 1-12).

The voltage across all branches of a parallel circuit is the same. This is so because all branches are connected across the voltage source. Current in a parallel circuit depends on the resistance of the branch. Ohm's law (discussed later) can be used to determine the current in each branch. You can find the total current for a

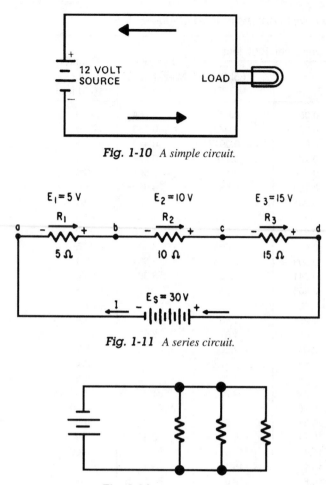

Fig. 1-10 A simple circuit.

Fig. 1-11 A series circuit.

Fig. 1-12 A parallel circuit.

Series-parallel circuits Series-parallel circuits are a combination of the two circuits. Figure 1-13 shows a series-parallel resistance circuit.

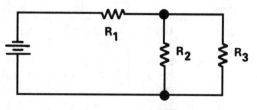

Fig. 1-13 A series-parallel circuit.

Open circuit An open circuit is one that does not have a complete path for electrons to follow. Such an incomplete path is usually brought about by a loose connection or the opening of a switch (Fig. 1-14).

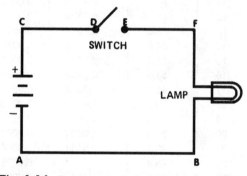

Fig. 1-14 An open circuit. Switch is in open position.

Short circuit A short circuit is one that has a path of low resistance to electron flow. It is usually created when a low-resistance wire is placed across a consuming device. The greater number of electrons will flow through the path of least resistance, rather than through the consuming device. A short circuit usually generates an excess current flow. This results in overheating, possibly causing a fire or other damage (Fig. 1-15).

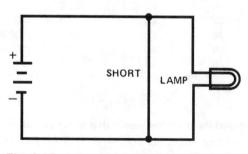

Fig. 1-15 A short circuit. The lamp will not glow.

parallel circuit by *adding the individual currents.* As a formula, this reads:

$$I_T = I_1 + I_2 + I_3 + \cdots$$

The total resistance of a parallel circuit *cannot* be found by adding the resistor values. Two formulas are used for finding parallel resistances. If there are *only two* resistors in parallel, a simple formula can be used:

$$R_T = \frac{R_1 \times R_2}{R_1 + R_2}$$

If there are more than two resistors in parallel, you can use the following formula. (You can also use this formula if there are only two resistors.)

$$\frac{1}{R_T} = \frac{1}{R_1} + \frac{1}{R_2} + \frac{1}{R_3} + \cdots$$

One thing should be kept in mind in parallel resistances: *The total resistance is always less than the smallest resistance.*

It is easy to compute the amount of current flowing in a circuit if the voltage and the resistance are known. The relationship between voltage, current, and resistance in any circuit is shown by *Ohm's law.*

OHM'S LAW

There are three basic quantities of electricity, and each has a relationship to the other two. A physicist named Georg S. Ohm discovered the relationship in 1827. He found that *in any circuit where the only opposition to the flow of electrons is resistance, there is a relationship between the values of voltage, current, and resistance.* The strength or intensity of the current is directly proportional to the voltage and inversely proportional to the resistance.

It is easier to work with Ohm's Law when it is expressed in a formula. In the formula, E represents emf, or voltage; I is the current, or intensity of electron flow; and R stands for resistance. The formula is $E = I \times R$. This is the formula to use to find the emf (voltage) when the current and resistance are known.

To find the current when the voltage and resistance are known, the formula to use is

$$I = \frac{E}{R}$$

To find the resistance when the voltage and current are known, the formula to use is

$$R = \frac{E}{I}$$

Using Ohm's Law

There are many times in electrical work when you will need to know Ohm's law. For example, you need to know Ohm's law to determine wire size in a particular circuit or to find the resistance in a circuit.

The best way to become accustomed to solving problems is to start with a simple problem such as the following:

1. If the voltage is given as 100 volts and the resistance is known to be 50 ohms, it is a simple problem and a practical application of Ohm's law to find the current in the circuit.

$$I = \frac{E}{R}$$

$$= \frac{100 \text{ volts}}{50 \text{ ohms}}$$

$$= 2 \text{ amperes}$$

2. If the current is given as 4 amperes (shown on an ammeter) and the voltage (read from a voltmeter) is 100 volts, it is easy to find the resistance.

$$R = \frac{E}{I}$$

$$= \frac{100 \text{ volts}}{4 \text{ amperes}}$$

$$= 25 \text{ ohms}$$

3. If the current is known to be 5 amperes and the resistance is measured (before current is applied to the circuit) and found to be 75 ohms, then it is possible to determine how much voltage is needed to cause the circuit to function properly.

$$E = I \times R$$

$$= 5 \text{ amperes} \times 75 \text{ ohms}$$

$$= 375 \text{ volts}$$

Figure 1-16 illustrates the way the formula works.

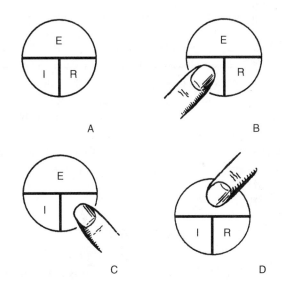

Fig. 1-16 *Ohm's law. Place a finger on the unknown value. The remaining two letters will give the formula to use for finding the unknown value.*

POWER

Power is defined as the rate at which work is done. It is expressed in metric measurement terms of *watts* (for power) and in *joules* (for energy or work). A watt is the power that gives rise to the production of energy at the rate of 1 joule per second (1 W = 1 J/s). A joule is the work done when the point of application of a force of one newton is displaced a distance of one meter in the direction of the force (1 J = 1 N × 1 m).

It has long been the practice in the United States to measure work in terms of *horsepower*. Electric motors are still rated in horsepower and probably will be for some time.

Power can be electrical or mechanical. When a mechanical force is used to lift a weight, work is done. The rate at which the weight is moved is called *power*. Horsepower is defined in terms of moving a certain weight over a certain distance in one minute. Energy is consumed in moving a weight, or work is done. The findings in this field have been equated with the same amount of work done by electrical energy. *It takes 746 watts of electrical power to equal 1 horsepower* (Table 1-3).

Table 1-3 Horsepower

One horsepower (1 hp) is usually defined as the amount of work required to move a 550-pound weight a distance of one foot in one second.

In most cases the *modern* way to measure power is in kilowatts rather than horsepower. In case a motor is specified in terms of horsepower, but is rated in watts or kilowatts, the conversion is simple:

1 horsepower = 746 watts

Divide the number of watts or kilowatts by 746 or 0.746, respectively, to find the horsepower rating.

However, it should be remembered that a 1-horsepower motor uses about 1100 watts to produce the power. The extra watts are consumed in heat and friction losses.

The horsepower rating of electric motors is obtained by taking the voltage and multiplying it by the current drawn under full load. This power is measured in watts. In other words, one volt times one ampere equals one watt. When put into a formula, it reads

Power = volts × amperes   or $P = E \times I$

Kilowatts

The prefix *kilo* means 1000. Thus, 1000 watts equals 1 kilowatt. The abbreviation for kilowatt is kW. There is also a unit known as the *kilowatthour*. It is abbreviated kWh and is equivalent to 1000 watts used for 1 hour. Electric bills are figured in kilowatthours. Usage for an entire month is computed on an hourly basis and then read in the kWh unit.

Power formulas are sometimes needed to figure the wattage of a circuit. Here are the three most commonly used formulas:

$$P = E \times I$$

$$P = \frac{E^2}{R}$$

$$P = I^2 \times R$$

This means that if any one of the three—voltage, current, or resistance—is missing, it is possible to find the missing quantity by using the relationship of the two known quantities. However, you must also know the power consumed. In later chapters you will encounter the problem of the I^2R losses and some other terminology related to the formulas just shown.

The milliwatt (mW) is sometimes used in referring to electrical equipment. For instance, the rating of the speaker in a transistor radio may be given as 100 mW or 300 mW. This means a 0.1-watt or 0.3-watt rating, since the prefix *milli* means one-thousandth. Transistor circuits are designated in milliwatts, but power-line electrical power is usually in kilowatts.

MEASURING ELECTRICITY

Electricity must be measured if it is to be sold, or if it is to be fully utilized. There are a number of ways to measure electricity. It can be measured in volts, amperes, or watts. The kilowatthour meter is the device most commonly used to measure power.

Meters

To measure anything, there must be a basic unit in which to measure. In electricity, the current (flow of electrons) is measured in a basic unit called the *ampere*. The current is usually measured with a permanent magnet and an electromagnet arranged to indicate the amperes. Such a device is necessary since we are unable to see an electron—even with the most powerful microscopes. Obviously, counting the number of electrons passing a given point in a second is impossible when there are no visible particles to count. Therefore, a magnetic field is used to measure the effect of the electrons.

The D'Arsonval meter movement uses a permanent magnet as a base over which a wire or electromagnet is pivoted and allowed to move freely. When current flows through the coil, a magnetic field is set up (Fig. 1-17a and b). The strength of the magnetic field determines how far the coil will be deflected. The polarity of the moving coil is the same as that of the permanent magnet. A repelling action results. This is in proportion to the strength of the magnetic field generated by the current flowing through the coil. The number of turns in the coil, times the current through the coil, determines the strength of the magnetic field. Since the meter coil is pivoted on jeweled bearings to reduce any friction, the movement is calibrated against a known source of current or against another meter. The scale on a new device is calibrated to read in amperes, milliamperes, or microamperes (Fig. 1-17c).

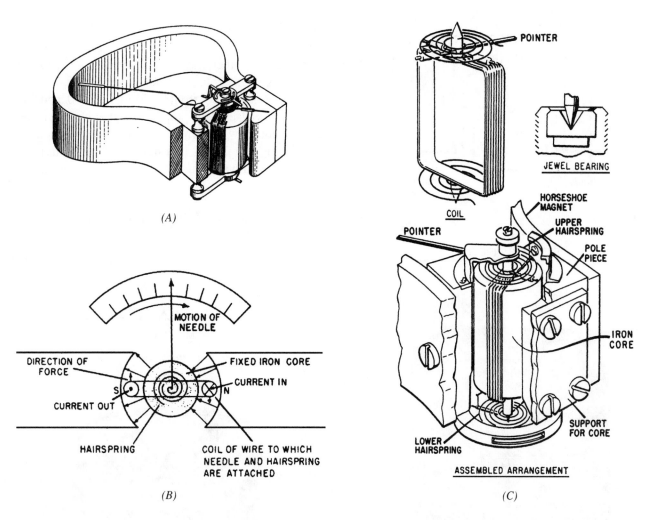

(A)

(B)

(C)

Fig. 1-17 *(A) D'Arsonval meter movement. (B) D'Arsonval meter movement showing completed unit in diagram form. (C) Assembled D'Arsonval meter movement.*

AC ammeter If alternating current (AC) is to be measured by a direct-current (DC) meter movement, a rectifier is inserted in the circuit (meter circuit) to change the AC to DC. It can then be measured by the meter movement. Otherwise the alternating current will make the needle on the meter vibrate rapidly. This vibration means there is little or no movement from zero (Fig. 1-18).

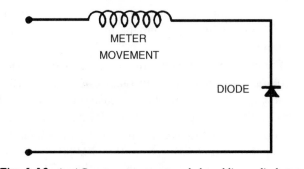

Fig. 1-18 *An AC meter movement made by adding a diode to a DC meter movement.*

Shunts A shunt is a means of bypassing current around a meter movement. A resistor of the proper size is inserted *across* the meter movement to bypass the current around the movement. Most of the current is bypassed. Only the necessary amount is left to cause the meter to deflect at its designed limit. The meter is calibrated to read on its scale the full amount of current flowing in the circuit (Fig. 1-19).

A number of shunts can be placed in a meter case and switched. A different resistor (shunt) can be switched in for each range needed. A meter with more than one range is called a *multimeter.*

A multimeter (one which can measure volts and ohms as well) is shown in Fig. 1-20. It can measure from 0 to 1 milliampere, 0 to 10 milliampere, and from 0.1 to 1 ampere. This is a 1-milliampere meter movement. It needs no shunt when used to measure as high as 1 milliampere. A shunt is switched in, however, to measure a range of 0 to 10 milliampere and 0.1 to 1 ampere. A switch on the meter can also add a diode to help with measuring AC.

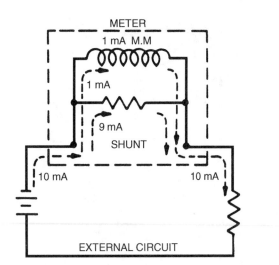

Fig. 1-19 A meter movement with a shunt to increase its range to 10 mA.

Fig. 1-20 A multimeter using a D'Arsonval meter movement.

across a power source (in parallel) (Fig. 1-21). A number of resistors, called *multipliers,* can be switched into a meter circuit to increase its range or make it capable of measuring higher voltages. The voltmeter in Fig. 1-22 is capable of measuring from 0 to 150, 0 to 300, and 0 to 750 volts by placing the proper multiplier into the meter circuit. Note how the terminals on top of the meter allow for placing test probes in a number of different positions. This allows the measurement range to be varied.

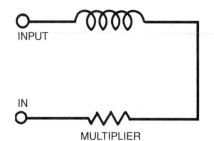

Fig. 1-21 A voltmeter's internal circuit. The multimeter extends the meter movement's range.

Fig. 1-22 DC voltmeter, laboratory model.

Common and *positive* holes have test leads inserted to attach to the circuit being measured. Common is negative (−) or black, and the positive is plus (+) or red. However, polarity isn't necessary to measure AC. Either lead can be used at any terminal in an AC circuit. *Ammeters are always placed in series in a circuit.* This usually means the circuit has to be broken and the meter inserted in the line.

Voltmeter The voltmeter measures electrical pressure, or volts. It is nothing more than an ammeter with a resistor added in the meter circuit. The high resistance of the voltmeter makes it possible to place it

Ohmmeter The basic unit for measuring resistance is the *ohm.* An ohmmeter is a device used to measure resistance or ohms (Ω). It is an ammeter (or milliammeter or micrometer) movement, modified to measure resistance (Fig. 1-23).

Figure 1-23 shows a multimeter capable of measuring ohms with three different ranges: $R \times 1$, $R \times 100$, and $R \times 10k$. This means it can measure from 0 to 200 ohms on the $R \times 1$ scale and 0 to 200,000 ohms on the $R \times 100$ range. Within the $R \times 10k$ or $R \times 10,000$ range, it is capable of measuring from 0 to 20,000,000 ohms or 0 to 20 megohms (*mega* means 1 million). The meter scale has to be multiplied by the 100 or 10,000 number

Fig. 1-23 A multimeter. This type is used to measure ohms, volts, and amperes.

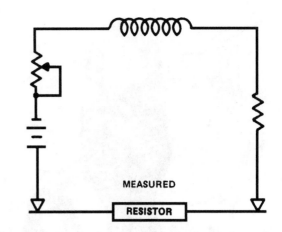

Fig. 1-25 Test leads are touching the leads of a resistor to measure its resistance.

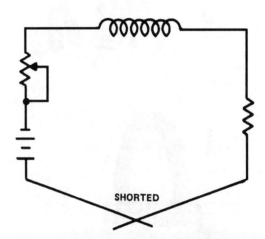

Fig. 1-26 Test leads are crossed (shorted) so resistor can be adjusted to make the meter read zero.

to have it read the proper value. By changing resistors, it is possible to vary the resistance measuring range of an ohmmeter. Figure 1-24 shows a basic ohmmeter using a 1-milliampere meter movement and its necessary parts. Note how the battery serves as the power source. This makes it necessary to turn off power whenever you read the resistance of a circuit. THE OHMMETER HAS ITS OWN POWER SOURCE. DO NOT CONNECT IT TO A LIVE CIRCUIT OR TO ONE WITH THE POWER ON. To do so will result in the destruction of the meter movement (Fig. 1-25).

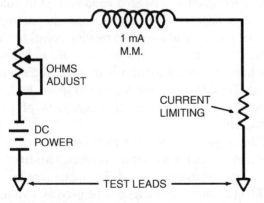

Fig. 1-24 The basic circuit of an ohmmeter.

Adjust the ohmmeter so that the meter reads zero before you start to use it to measure resistance (Fig. 1-26). This means you have adjusted the meter circuit to compensate for the battery voltage changes. Battery voltages decrease with shelf life. It doesn't matter whether the battery is used. It will, in time, lose its voltage.scale

Clamp-on meters Figures 1-27 and 1-28 show two different styles of an AC clamp-on meter. They are inserted over a wire carrying alternating current. The magnetic field around the wire induces a small amount of current in the meters. The scale is calibrated to read amperes or volts. Because the wire is run through the large loop extending past the meter movement, it is possible to read the AC voltage, or current, without removing the insulation from the wire. These meters are very useful for working with AC motors.

Wattmeter A wattmeter measures electrical power. Electrical power is figured by multiplying the voltage by the current. A wattmeter has electromagnetic coils (a coil with many turns of fine wire for voltage and a coil with a few turns of heavy wire for current). The voltage coil is connected across the incoming line. The current coil is inserted in series with one of the incoming wires. Two coils are stationary and in series

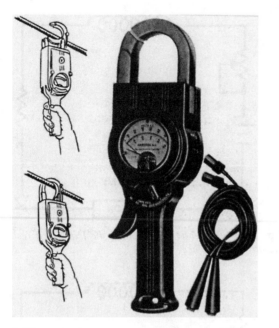

Fig. 1-27 *A clamp-on type of portable AC voltmeter-ammeter.*

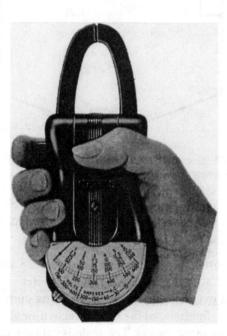

Fig. 1-28 *A clamp-on type of AC volt-ammeter.*

with a moving coil. The strength of the magnetic fields determines how much the moving coil is deflected. The deflection of the needle is read on a scale calibrated in watts. In this way the wattmeter measures the power consumed in 1 second.

Kilowatthour meter The kilowatthour meter was designed for measuring the electrical power used over a long time. The kilowatthour meter is often seen on the side of a house or other building. It measures power used over a certain time period, such as a month. A

kilowatthour meter measures the power consumed in terms of thousands of watts. Electric power is priced at a certain rate per kilowatthour.

The kilowatthour meter is a small induction motor. Meter torque is produced by an electromagnet called a *stator*, which has two sets of windings. One winding, called a *potential coil*, produces a magnetic field representing circuit voltage. Another winding, known as a *current coil*, produces a magnetic field that represents the load current. These two coils are arranged so that their magnetic fields create a force on the meter disk. This force is directly proportionate to the power, or watts, drawn by the connected load.

Permanent magnets are used to introduce a retarding, or braking force, which is proportionate to the speed of the disk. The magnetic strength of these retarding magnets regulates the disk speed for any given load so that each revolution of the disk always measures the same quantity of energy or watthours. Disk revolutions are converted to kilowatthours on the meter register.

Most meters are inserted into a socket on the wall of a structure. Removing the meter interrupts or terminates power without handling of dangerous high-voltage wires. Kilowatthour meters are tested by computers in the service centers of power companies (Fig. 1-29).

Other Types of Meters

There are other types of meters used to measure voltage and current. The D'Arsonval movement is only one of many types used today. The *taut band* type is basically the same as the D'Arsonval. However, a tightly stretched and twisted band is used to hold the coil and needle in place between the permanent magnet poles. In addition, no moving points touch the meter case. Thus, jeweled bearings are unnecessary. The band is twisted when it is inserted into the meter frame. Thus, it will cause the coil to spring back to its original resting place upon interruption of current through the coil.

- *Electrodynamometer.* The electrodynamometer type of meter uses no permanent magnet. Two fixed coils produce the magnetic field. The meter also uses two moving coils. This meter can be used as a voltmeter or an ammeter. It is not as sensitive as the D'Arsonval meter movement.

- *Inclined-coil iron-vane meter.* The inclined-coil iron-vane meter is used for measuring AC or DC where large amounts of current are present. This meter can be used as a voltmeter or ammeter.

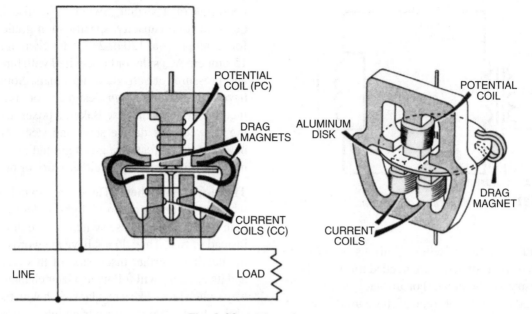

Fig. 1-29 *A kilowatthour meter.*

CONTROLLING ELECTRICITY

To make electricity useful, it is necessary to control it. You want it in the proper place at the proper time. Otherwise, it can do great damage—even kill. Electricity can be controlled by using switches, relays, or diodes. These devices are used to direct the current to the place where it will work for you. Each device is carefully chosen to do a specific job. For example, the relay is used for remote-control work. A diode is used to control large and small amounts of current in electrical as well as electronic equipment. A *diode* is a device that allows current to flow in one direction only. It can be used to change AC to DC.

Switches

There are a number of switches used for controlling electricity. Each switch has a different name. This helps designate it according to the job it performs. For instance, the single-pole, single-throw (SPST) switch is just that. It is a single pole that is moved either to *make connection* between two points or to *not make connection*. In the off position, the contacts are not touching, and the flow of electrons is interrupted.

The double-pole, double-throw (DPDT) switch can be used to control more than one circuit at the same time. It can be used to reverse the direction of rotation of a DC motor by reversing polarity.

The double-pole, single-throw (DPST) switch is used to control two circuits at the same time. It can be used as a simple on/off switch for two circuits. When it

is open, it interrupts the current in the two circuits. When it is closed, it completes the circuits for proper operation. This, too, is an open switch. It is meant to be used only on low voltages where the danger of shock is very much reduced.

The doorbell switch or door chime switch is extremely simple. It completes a circuit from the low-voltage transformer to the chime or bell (Fig. 1-30). When the button is pressed, it completes the circuit to the bell or chime, causing current to flow from the transformer to the chime (Fig. 1-31).

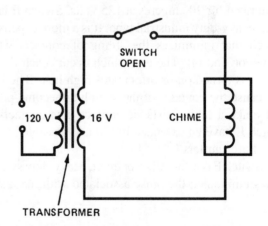

Fig. 1-30 *A doorbell or chime circuit. Switch is open.*

Toggle switches Toggle switches are used to turn various devices on and off, or to switch from one device to another. They are made in a number of configurations to aid in selection for a particular job.

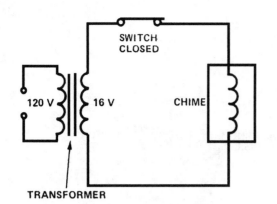

Fig. 1-31 *A doorbell or chime circuit. Switch is closed.*

Residential toggle switches Various shapes are encountered when switches are needed for use in business, industry, or the home. For instance, in Fig. 1-32 you will find examples of some of those used to switch 120 volts AC and 240 volts AC in common circuits used in lighting and small motors. In Fig. 1-32, switch A is a residential toggle switch, rated at 10 amperes on a 125-volt line. Note the absence of "plaster ears" near the long screws. Switch B is like switch A, except that it has wide plaster ears. These can be removed easily if not required to hold the switch rigid in its box. The plaster ears are scored (marked) so that they can be bent easily and removed. Switch C is more expensive than either A or B. It has *specification-grade* (top-of-the-line) quality with wide plaster ears. Note that the screws for attaching circuit wiring are located topside, instead of along the side of the switch.

Switch D is a high-capacity, heavy-duty, industrial type, rated for 20 amperes at 125 volts. Switch E is an extra heavy-duty industrial type. It is a more expensive switch that minimizes the arcing of contacts when turned on and off. The arc, which occurs each time a switch is turned on or off, creates high heat. The heat can cause the contacts of the switch to become pitted and make a high-resistance contact. The contacts in switch E have an extended life, made possible by the use of arc snuffers.

Switch F is a "no-klik," or quiet, heavy-duty switch. It has eliminated the noise associated with the on/off

operation of a switch. Switch G is also heavy-duty, quiet, of high capacity, specification grade, and good for 15 amperes at 120 or 277 volts. Switch H is a quiet 15-ampere AC, side- or back-wired with binding screw, and pressure, or screwless, terminals. Some switches have a *grounding strap* designed for use with non-metallic systems that use Bakelite boxes and for bonding between the device strap and steel boxes. These newer switches feature a green grounding-screw terminal. They are available in either ivory or brown.

Three-way switches Three-way switches are used where you need to control a light or device from more than one location. These switches have three terminals instead of two. They do not have the words *on* or *off* on the handles. Further discussion of this type of switch and its circuitry will follow in a later chapter. The three-way switch looks like a regular switch with the exception of the three terminals for wiring into a circuit.

Four-way switches Four-way switches are double-pole. They are used where a light or device needs to be controlled from three or more locations. If three controls are preferred, you need two three-way switches and one four-way switch. The four-way resembles the three-way, but has four terminals for connection into a circuit. It does not have *on* or *off* on its handle, since either up or down may be on or off, as you will see later. More information concerning the four-way switch will be given in a later chapter.

Other types of switches are available for different jobs of current control. They will be shown later, as they are introduced in connection with a specific job.

Switches are used to turn the flow of electricity on or off, thereby causing a device to operate or cease operation. Switches can be used to reverse polarity, and, as in the case of electric motors, the direction of rotation can be reversed by this action. Switches, as you have already seen, come in many shapes and sizes. The important thing to remember is to use a switch with proper voltage and current rating for the job to be done. A careful study of the types presented in this chapter will help in the proper selection of a switch for a particular job.

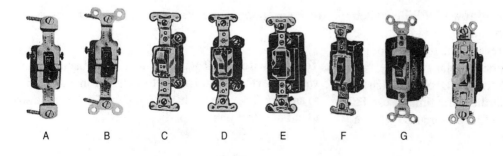

A   B   C   D   E   F   G

Fig. 1-32 *Switches used for residential wiring.*

Solenoids

Solenoids are devices that turn electricity, gas, oil, or water on and off. Solenoids can be used, for example, to turn the cold water on and the hot water off, to get a proper mix of warm water in a washing machine. To control the hot water solenoid, a thermostat is inserted in the circuit.

Figure 1-33 shows a solenoid for controlling natural gas flow in a hot-air furnace. Note how the coil is wound around the plunger. The plunger is the core of the solenoid. It has a tendency to be sucked into the coil whenever the coil is energized by current flowing through it. The electromagnetic effect causes the plunger to be attracted upward into the coil area. When the plunger is moved upward by the pull of the electromagnet, the soft disk (10) is also pulled upward, allowing gas to flow through the valve. This basic technique is used to control water, oil, gasoline, or any other liquid or gas.

The starter solenoid on an automobile uses a similar procedure. However, the plunger has electrical contacts on the end that complete the circuit from the battery to the starter. The solenoid uses low voltage (12 volts) and low current to energize the coil. The coil in turn sucks the plunger upward. The plunger then touches heavy-duty contacts that are designed to handle the 300 amperes needed to start a cold engine. In this way, low voltage and low current are used, from a remote location, to control low voltage and high current.

Solenoids are *electromagnets*. An electromagnet is composed of a coil of wire wound around a core of soft iron. When current flows through the coil, the core will become magnetized.

The magnetized core can be used to attract an armature and act as a magnetic circuit breaker (Fig. 1-34). (A circuit breaker, like a fuse, protects a circuit against short circuits and overloads.) In Fig. 1-34, the magnetic circuit breaker is connected in series with both the load circuit to be protected and the switch contact points. When excessive current flows in the circuit, a strong magnetic field in the electromagnet causes the armature to be attracted to the core. A spring attached to the armatures causes the switch contacts to open and break the circuit. The circuit breaker must be reset by hand to allow the circuit to operate properly again. If the overload is still present, the circuit breaker will "trip" again. It will continue to do so until the cause of the short circuit or overload is found and corrected.

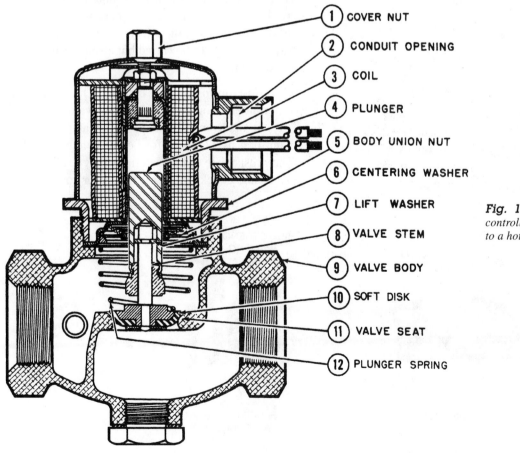

1. COVER NUT
2. CONDUIT OPENING
3. COIL
4. PLUNGER
5. BODY UNION NUT
6. CENTERING WASHER
7. LIFT WASHER
8. VALVE STEM
9. VALVE BODY
10. SOFT DISK
11. VALVE SEAT
12. PLUNGER SPRING

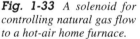

Fig. 1-33 A solenoid for controlling natural gas flow to a hot-air home furnace.

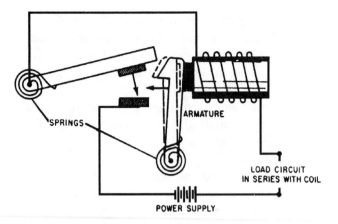

Fig. 1-34 *A magnetic circuit breaker.*

Relays

A relay is a device that can control current from a remote position through use of a separate circuit for its own power. Figure 1-35 shows a simple relay circuit.

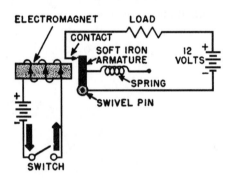

Fig. 1-35 *A simple relay circuit.*

When a switch is closed, current flows through the electromagnet, or coil, and energizes it. The pull of the electromagnet causes the soft iron armature to be attracted toward the electromagnet core. As the

armature moves toward the coil, it touches the contacts of the other circuits, thereby completing the circuit for the load. When a switch opens, the relay coil *de-energizes* and the spring pulls the armature back. This action breaks the contact and removes the load from the 12- volt battery. Relays are remote switches that can be controlled from almost any distance if the coil is properly wired to its power source.

Many types of relays are available. They are used in telephone circuits and in almost all automated electrical machinery.

RESISTORS

A resistor is a device used to provide a definite, required amount of opposition to current in a circuit. Resistance is the basis for the generation of heat. It is used in circuits to control the flow of electrons and to ensure that the proper voltage reaches a particular device.

Resistors are usually classified as either wire-wound or carbon-composition. The symbol for a resistor of either type is ⏦ .

Wire-wound resistors These are used to provide sufficient opposition to current flow to dissipate power of 5 watts or more. They are made of resistance wire (Fig. 1-36). Variable wire-wound resistors are also available for use in circuits where voltage is changed at various times (Fig. 1-37). Some variable resistors have the ability to be varied but also adjusted for a particular setting (Fig. 1-38).

Carbon-composition resistors These are usually found in electronic circuits of low wattage, since they are not made in sizes beyond 2 watts. They can be identified by three- or four-color bands around them. Their resistance can be determined by reading the color bands and checking the resistor color code.

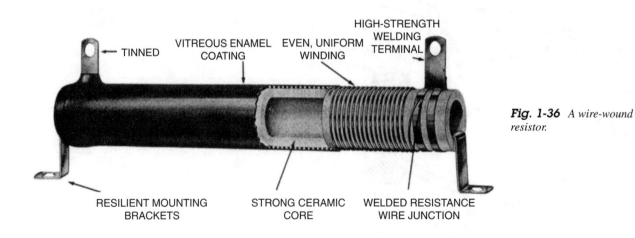

Fig. 1-36 *A wire-wound resistor.*

Fig. 1-37 *A variable wire-wound resistor. Note it has three terminals for connections.*

Fig. 1-38 *An adjustable wire-wound resistor.*

Resistor Color Code

0 Black	2 Red	4 Yellow	6 Blue	8 Gray
1 Brown	3 Orange	5 Green	7 Violet	9 White

The wattage ratings of carbon-composition resistors are determined by physical size. They come in ¼-watt, ½-watt, 1-watt, and 2-watt sizes. By examining them and becoming familiar with them through use, you should be able to identify wattage rating by sight. The larger the physical size of the resistor, the larger the wattage rating (Fig. 1-39).

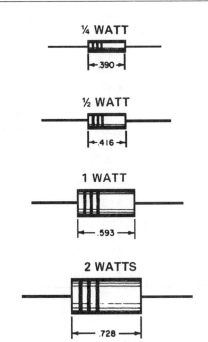

¼ WATT

←.390→

½ WATT

←.416→

1 WATT

←.593→

2 WATTS

←.728→

Fig. 1-39 *Carbon-composition resistors. Smaller resistors are now used in electronic devices since they use less power in their circuits.*

Resistor Color Code

Take a close look at a carbon-composition resistor. The bands should be to your left (Fig. 1-40). Read from left to right. The first band gives the first number according to the color code. In this case it is yellow, or 4. The second band gives the next number, which is violet, or 7. The third band represents the *multiplier* or *divisor*. If the third band is a color in the 0 to 9 range in the color code, it states the number of zeros to be added to the first two numbers. Orange is 3. Thus, the resistor in Fig. 1-40 has a value of 47,000-ohm resistance.

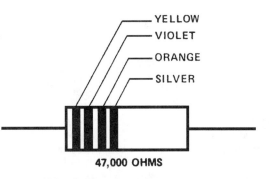

Fig. 1-40 *A 47,000-ohm resistor.*

If there is no fourth band, the resistor has a tolerance rating of ±20%, where ± means "plus or minus." If the fourth band is silver, the resistor has a tolerance of ±10%. If the fourth band is gold, the tolerance is ±5%.

Silver or gold may also be used for the third band. In this case, according to the color code, the first two numbers (obtained from the first two color bands) must be *divided* by 10 or 100. Silver means divide by 100; gold means divide by 10. For example, if the bands on a resistor are red, yellow, gold, and silver, the resistance is 24 divided by 10, or 2.4 ohms ±10% (Fig. 1-41).

Resistors are available in hundreds of sizes and shapes. Once you are familiar with electronics and electrical circuits, you will be able to identify various

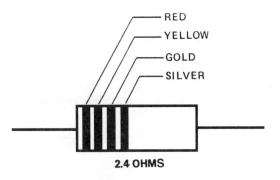

Fig. 1-41 *A 2.4-ohm resistor.*

components by their shape, size, or markings. Products for such circuits are constantly changing. New items are being marketed almost every day. To stay informed about these products, it is necessary to read the literature written about the industry. Each area of electrical energy has its own magazines to keep those on the job informed and up to date in their special fields of interest. One way to stay informed is to become familiar with the metric system of measurement. Table 1-4 illustrates how English units are converted to metric. Many of the measurements and values used in electricity and electronics are metric in origin.

Table 1-4 *Conversions*

To Convert	Into	Multiply by	Conversely, Multiply by
Inches	Centimeters	2.54	0.3937
Inches	Mils	1000	0.001
Joules	Foot-pounds	0.7376	1.356
Joules	Ergs	10^7	10^{-7}
Kilogram-calories	Kilojoules	4.186	0.2389
Kilograms	Pounds (avoirdupois)	2.205	0.4536
Kilograms per square meter	Pounds per square foot	0.2048	4.882
Kilometers	Feet	3281	3.048×10^{-4}
Kilowatthours	Btu	3413	2.93×10^{-4}
Kilowatthours	Foot-pounds	2.655×10^6	3.766×10^{-7}
Kilowatthours	Joules	3.6×10^6	2.778×10^{-7}
Kilowatthours	Kilogram-calories	860	1.163×10^{-3}
Kilowatthours	Kilogram-meters	3.671×10^5	2.724×10^{-6}
Liters	Cubic meters	0.001	1000
Liters	Cubic inches	61.02	1.639×10^{-2}
Liters	Gallons (liquid U.S.)	0.2642	3.785
Liters	Pints (liquid U.S.)	2.113	0.4732
Meters	Yards	1.094	0.9144
Meters per minute	Feet per minute	3.281	0.3048
Meters per minute	Kilometers per hour	0.06	16.67
Miles (nautical)	Kilometers	1.853	0.5396
Miles (statute)	Kilometers	1.609	0.6214
Miles per hour	Kilometers per minute	2.682×10^{-2}	37.28
Miles per hour	Feet per minute	88	1.136×10^{-2}
Miles per hour	Kilometers per hour	1.609	0.6214
Poundals	Dynes	1.383×10^4	7.233×10^{-5}
Poundals	Pounds (avoirdupois)	3.108×10^{-2}	32.17
Square inches	Circular mils	1.273×10^6	7.854×10^{-7}
Square inches	Square centimeters	6.452	0.155
Square feet	Square meters	9.29×10^{-2}	10.76
Square miles	Square yards	3.098×10^6	3.228×10^{-7}
Square miles	Square kilometers	2.59	0.3861
Square millimeters	Circular mils	1973	5.067×10^{-4}
Tons, short (avoirdupois 2000 lb)	Tonnes (1000 kg)	0.9072	1.102
Tons, long (avoirdupois 2240 lb)	Tonnes (1000 kg)	1.016	0.9842
Tons, long (avoirdupois 2240 lb)	Tons, short (avoirdupois 2000 lb)	1.120	0.8929
Watts	Btu per minute	5.689×10^{-2}	17.58
Watts	Ergs per second	10^7	10^{-7}
Watts	Pounds-feet per minute	44.26	2.26×10^{-2}
Watts	Horsepower (550 pounds-feet per second)	1.341×10^{-3}	745.7
Watts	Horsepower (metric) (542.5 pounds-feet per second)	1.36×10^{-3}	735.5
Watts	Kilogram-calories per minute	1.433×10^{-2}	69.77

Symbols	
Hertz	Hz
Kilohertz	kHz
Megahertz	MHz
RPM, rpm	r/min

Conversions	
1 mile	= 1.609344 kilometers
1 cubic foot (ft^3)	= 0.028316846592 cubic meters (m^3)
1 ton (short)	= 0.90718474 metric ton
1 foot	= 0.3048 meter

Cycles per second, cps, or ~ has been changed to *hertz*. The time interval of the *second* is part of the unit. So it is referred to simply as hertz. Instead of 60 cps, 60 cycles per second, or 60 ~, it is now 60 hertz or 60 Hz.

Revolutions per minute, rpm, or RPM, has been changed to r/min. This will probably be showing up later as the switch to metric becomes more noticeable.

Foot-pounds per minute and **foot-pounds per second** have been changed to pound-feet per second or pound-feet per minute.

2
CHAPTER

Producing Electricity

A S DISCUSSED IN CHAP. 1, THERE ARE A number of ways to produce electricity, some of which are commercially feasible. The use of magnetism is the most common method of generating electricity in large quantities.

Cells, or batteries, produce direct current (DC). A more economical way of producing DC, however, is with a mechanically driven generator. Mechanical force is used to rotate a wire loop in a magnetic field to generate electricity. The magnetic field is generated by a current-carrying wire, looped around a core (Fig. 2-1).

Since the current from the battery supply is DC with a negative (−) and a positive (+) polarity, the north (N) and south (S) fields are fixed in the positions indicated. If a conductor is moved in an upward direction through the magnetic field between the N and S poles (Fig. 2-1A), a current will flow in the conductor in the direction shown by the arrows. But if the conductor moves downward (as in Fig. 2-1B), then current flows in the opposite direction. The voltage generated will depend upon the intensity of the magnetic field, the number of turns of the wire, and the speed at which the wire passes through the magnetic field.

A simple generator is shown in Fig. 2-2. A loop of wire is wrapped around an iron core called an *armature.* Copper segments form the *commutator.* These segments

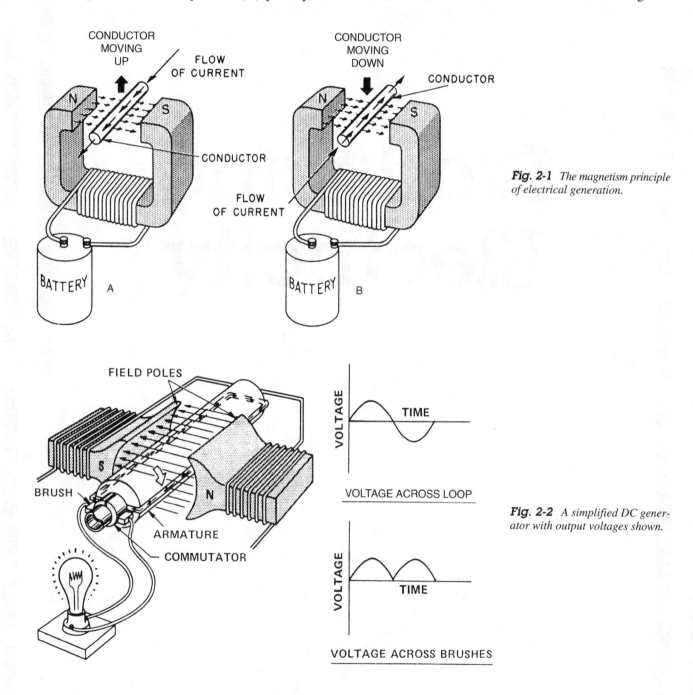

Fig. 2-1 *The magnetism principle of electrical generation.*

Fig. 2-2 *A simplified DC generator with output voltages shown.*

are attached to the ends of the loop of wire. They are insulated from the core and from each other. *Brushes* (conductors that make sliding contact) are so placed that they contact the commutator. They carry any generated electricity to the load, or consuming device. To produce electricity, the armature must be mounted between *field coils*. Thus, the magnetic force generated by the electromagnet will be cut by the rotating armature. Field coils form an electromagnet when they are wrapped around soft iron cores, known as *field poles*.

The position of the armature in Fig. 2-2 represents the point at which the armature loop cuts directly across the magnetic field. At this point it is generating its maximum output. When the armature has rotated another one-quarter turn, it will be moving parallel with the magnetic field. No output will be obtained from the generator. During the time that the armature rotates 360° (1 revolution), it generates a maximum and a minimum twice. It generates a maximum when passing across the S pole and again when crossing the N pole.

Voltage across the loop is represented as alternating current (AC) in Fig. 2-2. Voltage across the brushes is shown as pulsating direct current (PDC). The commutator acts as a reversing switch as the armature rotates in the different fields. As a result of the switching action, the current output is a series of maxima and minima with current flowing in only one direction.

An *AC generator* is usually referred to as an *alternator.* Alternators generate most of the electrical power used today. Large generating plants produce the electricity demanded by a world that has a constantly increasing need for energy.

Alternating current (AC) generators operate on the basic principle illustrated in Fig. 2-3. Atomic- and steam-powered generators use the same basic idea of electric power generation.

POWER PLANT OPERATIONS—AC

An alternating-current (batteryless) plant supplies electricity directly to the line to which electrical consuming devices are connected. It must be operated whenever any part of or all the equipment is to be used.

This type of plant supplies primary, or portable, power for field or construction work. It will operate electric lights, hand tools with universal motors, and other equipment and motors designed for AC. Alternating-current power has numerous other uses, especially for the home, farm, factory, industry, hospital, and communications fields. Its vital function in these fields is that of a reliable source of *emergency* electricity when there is an interruption in regular power service (Fig. 2-4).

Much of the equipment and most of the motors built today require alternating current. Thus, the AC plant is more widely used than any other type of generating plant.

An AC plant has other advantages. It is available in the widest range of sizes and with more types of control (ranging from manual start to full automatic). It will also operate a greater selection of low-cost equipment. Many appliances and some lighting, such as fluorescent, can be operated only on AC. Wiring distribution of the alternating-current system is probably the most flexible, efficient, and least expensive to operate.

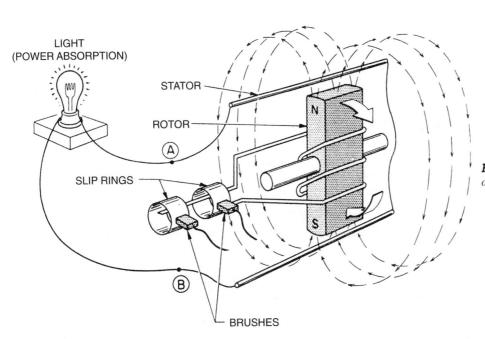

Fig. 2-3 *A simplified alternator, or AC generator.*

Fig. 2-4 *A diesel, water-cooled, six-cylinder, four-cycle engine. The larger sizes produce up to 500 kW.*

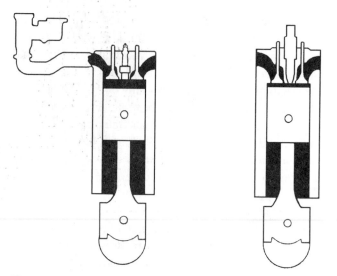

Fig. 2-5 *Gas and gasoline-type engine's intake on the left; diesel engine's intake on the right.*

The AC plant is ideal as a primary source of power. It is capable of operating a variety of large-capacity equipment on regular high-line power (overhead wires) when a conversion in power is made (transformers step down voltage). The plant may then be kept in reserve for emergency standby or other service. Such versatility makes the AC power plant unique.

An engine-driven AC plant with a DC starting system can supply a limited amount of DC power as well. It would be of 12-volt or 32-volt design, depending on the starting characteristics of the plant.

A plant must be operated whenever power is needed. There is an interruption in power when the plant needs repair or service. Such a problem is usually resolved through proper care and maintenance.

Mechanical Energy Sources

The previous discussion of electrical power plants covered their electric generator component. Now let us review the various types of generator driving power that are often used. The prime source of mechanical power for driving the generator is usually one of these three types: gasoline, diesel, or gas engine (Fig. 2-5).

A high degree of efficiency has been reached in the design and production of these internal combustion engines. This ensures that any one of the three types will produce a dependable source of power.

All engines must maintain a specific generator speed to develop the desired frequency. For example, 60 hertz (Hz) requires 3600 rpm, or some fraction ($\frac{1}{2}$ or $\frac{1}{3}$) of 3600, which would be 1800 or 1200. For 50 hertz, this must be 3000 rpm or 1500 or 1000.

Gasoline engine The gasoline engine is probably best known, since its basic design is the same as that of the en-

gine used in automobiles. Sometimes, however, there may be a low requirement for power in some plants. Here a gasoline engine may need only one or two cylinders, instead of four or more, as required in an automobile.

Most of the engines in power plants are of the four-cycle type that operate at either 1800 or 3600 rpm. Some are two-cycle engines, operating at 3600 rpm. These are used for extremely lightweight, portable units. Because these engines present certain service problems, the trend in design is toward four-cycle engines that run at moderately high speeds. By doing this, more generating capacity is obtained, with less plant weight. Older engines ran at 300 or 450 rpm. Modern engine design has increased the speeds. Now, 1800 rpm is established as standard, and 3600 rpm is universally accepted (Fig. 2-6).

All plant engines have compression ratios designed to take full advantage of high-octane gasoline. This gives greater operating efficiency.

Diesel engine The diesel engine is also of the internal combustion type. It uses the heat of compressed air to ignite the fuel, instead of an electric spark, as the gasoline and gas engines do. Production problems, very high initial costs, and limited use prevented early customer acceptance of diesel. Today, diesels are in demand. An important reason for their widespread use is their ability to run on comparatively low-cost fuel, their low operating costs, and their long life.

Two combustion cycles are available in diesel engines: two-cycle and four-cycle. Both types are practical. However, four-cycle engines are sometimes more economical. They use less fuel than two-cycle engines.

Due to extremely high combustion pressures, all components used in this engine must be larger than those of gasoline engines. They must be made of very

Fig. 2-6 *This gasoline engine produces 13 hp at 1800 rpm. It has remote start/stop capability with 12 volts DC. The generator puts out 120 volts, 1ϕ, 60 hertz, at 4 kW. Water-cooled model is shown here.*

strong material. For example, the injection pump times and measures the amount of fuel for each cylinder. This part must have very close tolerances for good starting and efficient operation. This adds to the expense of the engine (Fig. 2-7).

Gas engine Natural and manufactured gas (or a combination of both) and bottled, or liquefied petroleum

Fig. 2-7 *A diesel engine—single-cylinder, four-cycle, 5.7-horsepower, air-cooled. The generator puts out 3 kW at 120/240 volts, 50 hertz, 1ϕ, AC. This engine can be used for emergency home supply.*

(LP), gas are widely used as engine fuels. A standard gasoline engine, equipped with special carburetion, permits the use of such fuel. In theory, combustion is the same, except the fuel is already in a gaseous state. Thus, the process of gasoline vaporization is eliminated. Use of gas fuels allows the engine to run cleaner, with lower maintenance costs as a result.

Natural gas has a high enough British thermal unit (Btu) content [equivalent in the SI metric system is the joule (J)] to permit the engine to develop near-rated horsepower. Both manufactured gas and a mixture of it and natural gas have a reduced Btu content. This lowers the output of the engine and plant. The plant may require *derating* if the fuel used has a Btu content of 1000 or less. Derating means lowering the plant output because the fuel does not contain enough energy to meet the standard fuel value.

Bottled, or LP, gas is usually *butane* or *propane* or a mixture of both. It is supplied according to local and climatic conditions. It is stored in tanks in liquid form under pressure. When pressure is reduced, the liquid becomes a gas and operates the same as natural gas. In all cases, the Btu content is high. The engine output is comparable to that obtained from gasoline.

Economical Considerations of Various Engines

Because of the apparent lower operating costs of diesel engines using oil for fuel, the person responsible for ordering plant machinery might be inclined to specify that type of engine. There are, however, several factors to consider.

Original cost Diesel plants cost more than either gasoline or gas engines for several reasons. There is not the high mass production, which would help reduce manufacturing costs. More expensive components are required for diesel engines. Because they are heavier than other engines, transportation costs are higher for them. Gas plants are slightly higher in cost than gasoline plants.

Installation cost For an average job, installation costs are about the same for the three types of plants. But when a gasoline plant must meet local fire codes for a building installation, the cost of gasoline plants may be the highest of the three plant types. Strict regulatory codes exist because volatile fuel is being stored within the building and adequate protection must be provided. Installation of gas engine plants is probably the least expensive, with the diesel rating next in economy. The same comparison would probably apply to insurance on a building. Check insurance policies to make sure, however.

Starting methods Gasoline and gas engine plants are easier to start and are manufactured with more types of starting systems than diesel plants (Fig. 2-8). Diesels can be hand-cranked, generally in sizes up to 3 kilowatts. Other types can be hand-cranked in all sizes. All plants can be electrically started. Smaller gasoline and gas plants may be started through the exciter winding of the generator. Larger sizes may be started by an automotive-type starter. Only small diesel plants can be exciter-cranked. All diesels can be electrically cranked, however. The ambient temperature (air temperature surrounding the plant) must be above 0°F to ensure starting. This may be accomplished with glow plugs, air heaters, or room and oil-base heaters.

There are many types of starting methods and variations of those methods. Let us consider the five most common methods:

1. *Manual.* This plant, using magneto (permanent magnets) ignition, is manually started by hand crank or rope. It is stopped by a push button. This method is used to start portable, mobile, intermittent, and sometimes, emergency-service units (Fig. 2-8).

2. *Electric.* This plant, using magneto or battery ignition, is electrically started or stopped by means of push buttons on the unit. Stopping is possible from any push button located up to 250 feet (80 meters) from the plant. Storage batteries furnish current for cranking (Fig. 2-8).

3. *Remote control.* This type of plant is electrically started or stopped by push buttons mounted on the plant, or from any number of push-button stations located up to 250 feet (80 meters) from the plant (Fig. 2-8). A set of automotive-type batteries supplies the electric power for cranking. These batteries are kept charged by a special circuit, controlled at the plant. Depending on the voltage and capacity of the batteries, the starting circuit may also be used to supply a small amount of DC for lighting or other purposes.

The remote control is practical for many installations. When used with many start/stop stations, it provides almost the same service as an automatic plant. However, its operations are controlled. This eliminates haphazard operation and results in greater efficiency. A water pump cooling system may also be run with this method. However, automatic heating or refrigeration requires a full-automatic starting method. To convert from remote control, all that is necessary for most AC remote control plants is the addition of a full-automatic control panel designed for this purpose.

4. *Full automatic.* This plant starts automatically whenever an electric lighting load of about 75 watts (resistance load much higher) is turned on. It continues to run until the load is turned off. This starting method will operate a refrigerator, or any properly connected automatic equipment, without any individual attention. Basically, this method uses a standard remote-control plant plus an automatic control that makes the plant fully automatic in starting. Thus, a remote-control plant may be converted at any time to full-automatic starting by the addition of the panel.

A plant with this automatic starting method is likely to run more frequently than a plant with other types of starting. Any minimum electrical load will start it.

Special consideration must be given to the size and use of a full-automatic plant. If several automatic devices are to be used, with a possibility that they may all operate at one time, the plant must be large enough to handle the total starting and running load. A plant using the other starting methods may be smaller in size because its load and usage can be controlled.

5. *Automatic load transfer.* This starting method is designed for emergency standby plants (Fig. 2-8). Whenever there is a power interruption on a commercial line, an automatic transfer disconnects the load from the line. It starts a standby engine and keeps the plant running. It connects the load to the plant, then reverses these procedures when normal power is resumed. It ordinarily supplies a trickle charge to the starting battery when the plant is not running.

An automatic load transfer may be installed and connected to any AC remote-control plant to provide this type of starting. The plant and transfer mechanism may be connected in several ways. One method is to allow the combination of the two to handle the entire load. Another solution provides that the plant and transfer handle only a certain portion of the load on a preselected circuit. The size of

Fig. 2-8 *Starting methods for power plants. Left to right: hand crank, switch remote-controlled, and automatic load transfer.*

both plant and transfer is selected according to electrical requirements.

Certain relays may be added to the transfer to service special needs as required. These include voltage-sensitive relays and time-delay relays. Voltage-sensitive relays will start the plant when line voltage varies above or below a predetermined value. Time-delay relays delay starting and stopping. The plant does not start on momentary power outages or shut down before the high-line power is securely restored. Where the starting load is larger than plant capacity, time-delay relays are necessary to transfer the load to the plant in stages so that motors can be started one at a time.

Operation The main point to consider is the cost and availability of the fuels. In almost all cases, diesel fuel costs are the lowest, with gasoline and gas being about the same for similar Btu values. Prices, of course, are subject to change. Exceptions in fuel economy occur where a plant has access to a free gas well. This would obviously give it the lowest fuel cost when compared to other types of plants. However, if this plant operates only a few hours per year, the savings on fuel costs would not offset higher initial equipment costs.

Availability of fuels is of prime importance. A standard gasoline at somewhat stable prices is available almost everywhere. However, the other fuels—gas and diesel—cannot always be secured. This should be a determining factor in selection of the type of engine for a plant.

Lubricating oil used in the engine crankcase may be slightly more expensive per quart for the diesel plant. The small amount required, as compared to the operating fuel, does not make it an important price factor.

Gasoline and gas plants require about equal expense for engine maintenance. Such costs may be lower than for the diesel engine, even though the latter has fewer points to service. An operator of a gasoline, gas, or diesel plant can, in most cases, take care of his or her own maintenance. However some diesel engines may require special training for servicing and maintenance personnel. This type of specialized service has become more readily available. It does add to the overall cost of diesel plant operations.

Overhaul All engines require a periodic overhaul, with carbon removal and valve grind. In addition, certain parts require replacement, depending on their type and length of service. A complete overhaul for gasoline and gas plants will probably cost less than for diesel plants. (Such an overhaul, however, will be required more often than for the diesel plants.) As a result, total overhaul costs for the life of a diesel plant may be

lower than for other types because most service to a diesel plant is usually for accessory equipment.

Engine cooling Two basic methods are used for the job of cooling an engine that develops a large amount of heat in its operation: *air* and *water* (Fig. 2-9).

Water-cooled engines have radiators that are supplied by municipal water. This usually raises costs higher than for air-cooled systems. The single-duct radiator heat exhausting system has been equaled by new air-cooled systems.

It may be necessary someday for you to operate, or at lest identify, these power plants. The information presented should be sufficient for you to detect advantages and disadvantages. This information might enable you to recommend a particular unit.

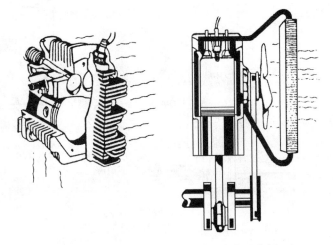

Fig. 2-9 *Air-cooled (left) and water-cooled (right) engines.*

FALLING-WATER GENERATORS

The principle of electricity generation is the same, whether it uses mechanical energy produced by gasoline or diesel engines, or falling-water or nuclear energy. In this part of the chapter, you will learn about the production of electricity by falling water. This method uses the energy released by the force of water falling over a distance to turn a generator and produce electricity.

Niagara Power Project

The Niagara Power Project is an example of falling water used to generate huge amounts of electrical energy. As the description of the operation and utilization of the plant progresses, you will probably realize the similarity of this type of power generation to that of fossil-fuel and nuclear power plants. Although these different types of generators of electricity are the same, the method for generating mechanical energy is different. (Mechanical energy is needed to turn the shaft upon which a coil of wire is revolved in a magnetic field.) Initial cost of building a power plant of this magnitude is extremely high. However, the system is comparatively maintenance-free for about 100 years.

Niagara is one of the largest hydroelectric developments in the Western world. Total installed power at Niagara's two generating plants (Robert Moses and Lewiston) is 2,190,000 kilowatts (Fig. 2-10). There are

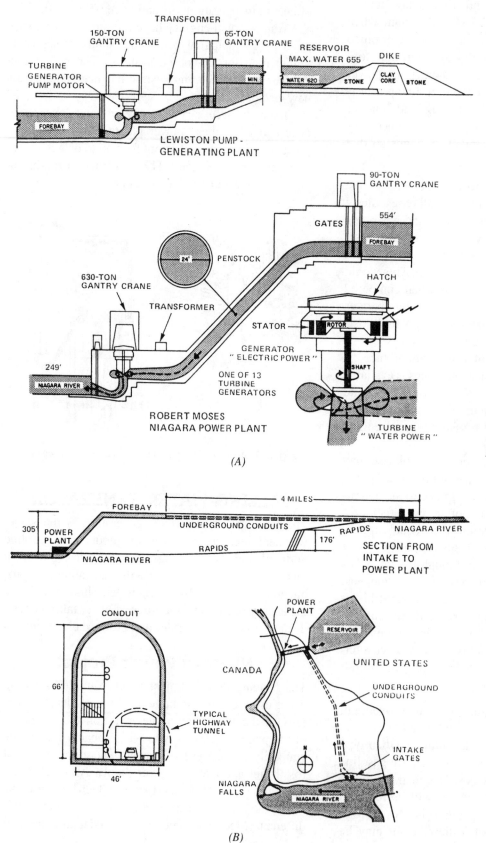

Fig. 2-10 (A) Hydroelectric generating plants at Niagara Falls, New York. (B) Robert Moses power plant at Niagara Falls.

340 miles (544 kilometers) of transmission lines to interconnect with the Niagara Power Project and the dam at St. Lawrence.

Switchyard

The Niagara switchyard is situated on a 35-acre (14-hectare) site south of the power canal, halfway between the Robert Moses and the Lewiston power plants. The purpose of the switchyard is to collect and meter power from the generators and send it out over transmission lines. The switchyard has three voltage sections: 115 kV, 230 kV, and 345 kV.

Power enters the switchyard from the Robert Moses plant through seven 115-kV cable circuits and six 230-kV cable circuits. It enters from the Lewiston plant by means of four 230-kV circuits. The cables are installed in underground power tunnels.

NUCLEAR POWER PLANTS

A nuclear power plant is in many aspects similar to a conventional fossil-fuel-burning plant. The chief difference is in the way the heat is generated, controlled, and used to produce steam to turn the turbine generator (Fig. 2-11).

In a nuclear power plant, the furnace for burning coal, oil, or gas is replaced by a reactor that contains a core of nuclear fuel. Energy is produced in the reactor by a process called *fission*. This fission process splits the center, or nucleus, of certain atoms when they are struck by a subatomic particle called a neutron. The resulting fragments, or fission products, then fly apart at great speed. They generate heat as they collide with surrounding matter.

The splitting of an atomic nucleus into parts is accompanied by the emission of high-energy electromagnetic radiation and the release of additional neutrons. The released neutrons may in turn strike other fissionable nuclei in the fuel, causing further fissions.

A nuclear reactor is thus a device for starting and controlling a self-sustaining fission reaction. The nuclear core of the reactor generally consists of fuel elements in a chemical form of uranium and thorium, or plutonium, depending on the type of reactor. Heat energy is produced by the fissioning of the nuclear fuel. A coolant is then used to remove this heat energy from the reactor core so that it can be used in producing electricity.

Water-Cooled Reactor

In a water-cooled reactor, the nuclear fuel, which has been compacted into uniform pellets, is placed into tubes or fuel elements. These fuel elements are then sealed at the top and bottom. They are arranged by spacer devices into bundles called fuel assemblies. The spacer devices separate the fuel elements to permit coolant to flow around all the elements. This process is necessary to remove heat produced by the fissioning uranium atoms. Scores of fuel assembles, precisely arranged, are required to make up the core of a reactor (Fig. 2-12).

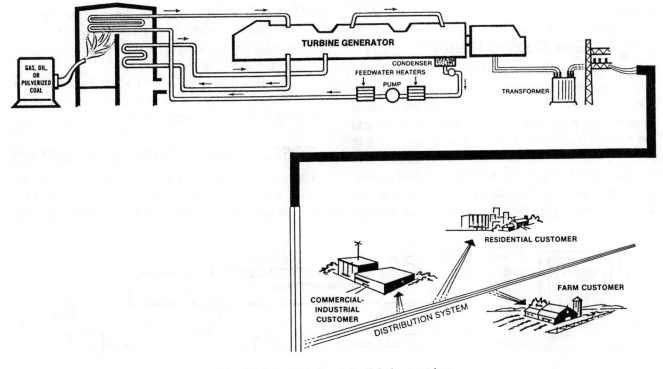

Fig. 2-11 *A conventional fossil-fuel power plant.*

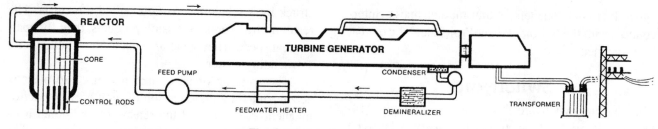

Fig. 2-12 *A boiling water reactor.*

This geometric arrangement is necessary for several reasons. Nuclear fuel, unlike fossil fuel, has a very high energy density; that is, tremendous quantities of heat are produced by a small amount of fuel. Therefore, the fuel must be arranged as stated above to permit the coolant to carry away the heat. To accomplish this, the fuel must be dispersed rather than lumped together in a large mass.

Chemical reactions between the fuel and the coolant must be avoided. Thus, as a safety precaution, radioactive materials produced must be enclosed. For these reasons, the fuel is contained in individual tubes, or fuel elements. The cladding material, from which the tubes are made, must meet rigid specifications. It must have good heat-transfer characteristics. It must not react chemically with either the fuel or the coolant. Also, it must not absorb the neutrons produced in the fissioning of the fuel at a rate that could be detrimental to the chain reaction. The cladding material generally used for this purpose is thin-walled stainless steel. An alloy of the element zirconium can also be used.

Gas-Cooled Reactor

A gas-cooled reactor, which utilizes the inert gas helium as the coolant, has a different core structure than the water-cooled reactor. The fuel elements are made of graphite. Graphite acts as the structural material and the neutron moderator. It also serves as the cladding material. Nuclear fuel consists of both uranium and thorium. It is formed into the center of the fuel element. Because the helium is inert, the graphite serves as cladding for the nuclear fuels. (An inert gas

will not react with or corrode the graphite or any other structural material.)

Physically, the fuel elements for a gas-cooled reactor are much larger than those of a water-cooled reactor. They are not bundled into individually arranged fuel assemblies and spaced for the circulation of coolant, as are those in water- cooled systems. Despite their larger size, however, several hundred fuel elements are required to make up the core of a reactor (Fig. 2-13).

Nuclear fuel, whether in the form of fuel assemblies for a water-cooled reactor or fuel elements for a gas-cooled reactor, is placed in the reactor to produce heat. The heat in turn is converted to electricity. After several years of operation, a nuclear core must be replaced. With the passage of time, the absorption of neutrons by the accumulated fission products is so great that there are too few neutrons remaining to maintain a chain reaction (Fig. 2-14).

After fuel assemblies, or elements, are removed from the reactor, they retain some material. The material must be reclaimed because of its economic value. In fact, only about 1 to 3% of the uranium has been "burned" in nuclear reaction. The remaining 97 to 99% is locked in the hundreds of fuel elements of the "spent" core (Figs. 2-15 and 2-16).

The procedure by which the nuclear fuel is reclaimed is called *reprocessing*. The first industry-owned reprocessing plant began operation in 1966.

FOSSIL-FUEL POWER GENERATORS

The stream needed for driving turbines (which in turn drive electric generators) must be produced by heat. The method of heat production often becomes a rather

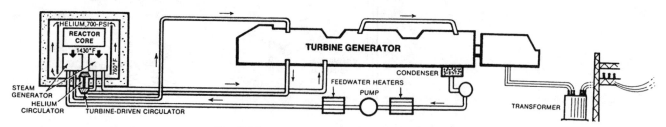

Fig. 2-13 *A high-temperature, gas-cooled reactor.*

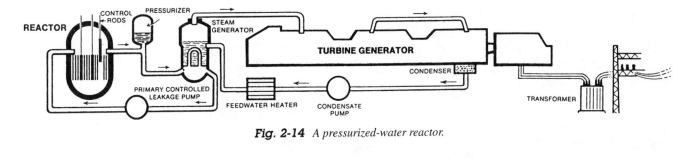

Fig. 2-14 *A pressurized-water reactor.*

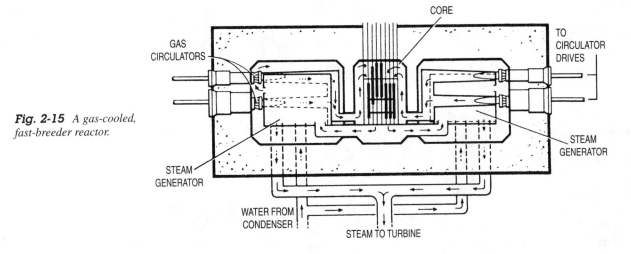

Fig. 2-15 *A gas-cooled, fast-breeder reactor.*

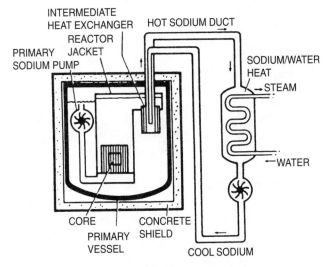

Fig. 2-16 *A liquid-metal, fast-breeder reactor.*

difficult engineering problem. With the development of some dependable sources of heat from fossil fuels, former design problems have been simplified. Almost any substance may be used as a fuel. If it can be pulverized and fed into furnaces with extremely high temperatures, it will burn.

Coal, Natural Gas, and Oil

Coal has been one of the most abundant sources of energy. It is used frequently to produce electricity by providing steam in those power plants designed for the use of coal. It is not, however, the only fossil fuel used to produce steam for power plants. Natural gas and oil are also used. There are some problems associated with these fuels. The waste products can pollute the atmosphere. Combustion products discharged into the atmosphere from a steam generator must meet Environmental Protection Agency (EPA) emission standards for particulates, sulfur oxides, and nitrogen oxides. *Clean fuels* for a steam generator may be defined as those fuels whose combustion products do not require stack gas cleanup to meet the limits set by antipollution agencies.

Synthetic gas Steam generation from synthetic, low- or medium-Btu gas is a development probably several years in the future. This is due to the lack of availability of a commercial gasifier system at this time. Conversions of existing units to gas must be handled on an individual basis, to ensure operation within original design limitations.

Coal-fired power plants Figure 2-17 shows a Niagara Mohawk power plant in the town of Tonawanda, New York, located on the Niagara River. A huge pile of coal is needed to fire the furnaces. The river here also provides an ample supply of water. Figure 2-18 shows the inside of the power plant. Here the turbines, fed by steam, rotate electrical generators to produce the power needed.

Fig. 2-17 Niagara Mohawk's coal-fired plant located on the Niagara River supplies power for Buffalo, New York.

Fig. 2-18 Generator room where steam turbines drive the alternators to produce electricity.

Figure 2-19 shows an operator in the control center of a large nuclear power plant. The control centers of coal-powered plants are somewhat similar in appearance. In both, operation is by remote control. Monitoring charts are checked constantly for performance and demand.

Fig. 2-19 The power control center in a large nuclear plant.

CHAPTER

Distributing Electricity

URBAN AND SUBURBAN DISTRIBUTION

Electricity may be generated by fossil-fuel power generators, by atomic-powered generators, or by the use of falling water to drive generators. It is also possible to generate power by engine-driven generators. In Alaska, for example, long distribution lines are impractical due to ice conditions that would result in line damage during storms. In Alaska, engine-driven generators are common.

Once electricity is generated in sufficient amounts for consumption in large quantities, the second necessary step is to get the energy to the consumer. Herein lies a distribution problem: that of stringing and maintaining long lines.

This chapter is concerned with some of the problems and methods of distributing electricity after it is generated by public utilities.

Figure 3-1 diagrams the process of getting power from the generating plant along high-voltage transmission lines, to voltage step-down substations, and then to office buildings, stores, apartments, and large factories. Further reductions are necessary in voltage to reduce the power to proper voltages (120-240) for home use.

The main concern of most utility companies is to get the greatest possible amount of usable energy to customers. Voltage losses on the way from the plant to consumers are minimized by proper use of step-up and step-down transformers and the correct size of wire.

Electrical Power

Most electrical power is generated as three-phase (3ϕ). It is stepped up to 132,000 volts, 238,000 volts, or even 750,000 volts. The frequency is 60 hertz. Sometimes, however, 25 hertz AC is generated for use by some consumers who still may have older equipment that uses this frequency. Motors, and other equipment using this lower frequency, would be too expensive to replace. However, most 25-hertz equipment, when worn out, is replaced with 60-hertz equipment. A separate distribution system is necessary to send 25-hertz power to its destination. This is expensive.

U.S. power companies In the United States, power is generated by shareholder-owned electric companies, by government-owned facilities such as the Tennessee Valley Authority (TVA), and by privately owned companies. The most popular system in the United States

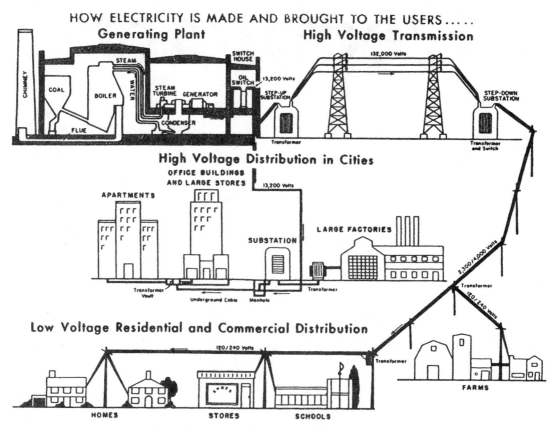

Fig. 3-1 *The generation and distribution of electricity.*

is the shareholder-owned. In this system, an electric company is owned by people who buy shares in the company on the stock exchange.

Shareholder-owned electric companies differ in two major respects from commercial and manufacturing companies:

- Their activities are confined to geographic service areas established by regulatory agencies.
- They operate free of competition from other electric companies as long as the best interests of the public are served.

Competition by suppliers of other forms of energy, such as gas, coal, and oil, is intensive, however.

Electric utilities are granted exclusive franchise areas because of the extremely large investment required of them to supply electricity, compared to the investment necessary for other types of businesses.

Canadian power companies In Canada the power companies are owned and controlled by the provinces. Because equipment standards are set by the provinces, companies can be very particular concerning the equipment used on their lines. In addition, they can specify and regulate inspection of devices hooked to the power lines (such as home appliances). In this way the power companies, along with the Canadian Standards Association, are able to demand and get safer manufactured devices in homes and industry. Their regulations can specify whether equipment is safe to operate. The Canadian companies can also designate the type of wiring for a house, business, or industry.

Fig. 3-2 *The distribution network for a large generating plant.*

High-voltage transmission Figures 3-2 through 3-5 illustrate some of the devices and equipment used to get the high voltage from the generator to the customer.

Fig. 3-3 *The control room of a large generating plant.*

Fig. 3-4 *Installing a new 238,000-volt transmission line.*

Fig. 3-5 *Note the size of the transformers in comparison to the man nearby. This is a high-voltage substation.*

Local Distribution

Figures 3-6 through 3-15 are concerned with transferring power from the substation to the local user. Most of these photos show overhead installations of power distribution systems. There are also a number of underground installations. Large cities and most suburban areas are now demanding removal of unsightly wires and poles. Underground wiring, although more expensive, is often used.

Underground systems Underground systems for distributing electrical energy are in the middle of a revolution that began about 10 years ago. A prime objective has been to reduce the cost of these systems so that they could be used in areas previously serviced only by overhead systems. Aluminum, long known as an economical electrical conductor, has contributed to this revolution.

Actually, underground distribution is as old as the electric power industry itself. In 1882, Thomas A. Edison built the first central power station on Pearl Street in New York City. He installed feeders and mains under the city streets to distribute the power generated in the station to its customers. These were pipe-type cables with a pair of copper conductors that were separated from each other, and from the enclosing pipe, by specially designed washers. The pipe was then filled with an insulating compound. This compound consisted of asphaltum boiled in oxidized linseed oil, with paraffin and a little beeswax added.

The development of lead-sheathed insulated cables that could be pulled into previously installed ducts greatly simplified construction of underground distribution systems. However, the high costs of such systems remained. As a result, for many years underground distribution was limited largely to the central portions of cities. Here load densities are high, and the congestion that would result from overhead systems is highly undesirable (Fig. 3-6).

Today, costs of underground distribution have been reduced to levels that are economically practical for service to light- and medium-density load areas. Underground systems now are being installed in most new residential areas. Home buyers in these areas generally agree that the improved appearance and enhanced value of their property more than justify the added cost of underground service (Figs. 3-9 and 3-10).

Some utilities are replacing existing overhead systems with underground systems in residential areas, particularly where load growth requires added capacity. Even some rural systems are being installed underground today.

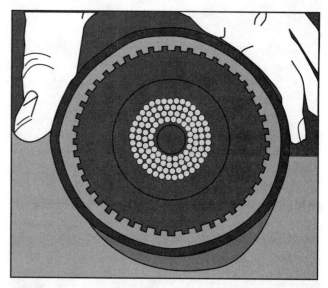

Fig. 3-6 *Cross-sectional view of a high-voltage underground cable. Note the various materials used for insulation.*

Fig. 3-7 *Stringing new cables for local service.*

Fig. 3-8 *Providing telephone service to an apartment complex. In most neighborhoods, telephone and electrical power use the same poles.*

Fig. 3-9 *Laying wires in a trench for an underground distribution system. These wires will be used to provide underground power circuits for homes in a subdivision.*

Reduced costs of underground systems have been achieved through the development of efficient new materials and low-cost system components. Progress has also been made in simplifying construction techniques and installation methods. One of the most important developments has been new synthetic insulating material, notably thermosetting, or cross-linked polyethylene. Cables with this type of insulation may be buried directly in the earth without expensive extra ductwork. The development of improved cable plowing and trenching equipment has also helped to reduce installation costs.

RURAL ELECTRICITY

The industrial revolution of the nineteenth century, which had transformed living styles in cities the world over, scarcely touched life on the farms of the United States. As a result U.S. farmers at the dawn of this present century were earning a living in a way that had changed little from that of the first colonists who settled along the Atlantic seaboard. The tools used were

Fig. 3-10 *A buried transformer is spliced into service. Notice how it is buried in an enclosure.*

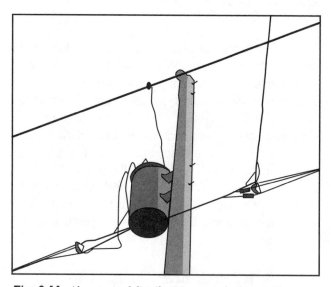

Fig. 3-11 *Aboveground distribution system for homes. The transformer is connected to the top wire for high-voltage input. The three wires below the transformer are the low-voltage (240-volt) output that goes to the house.*

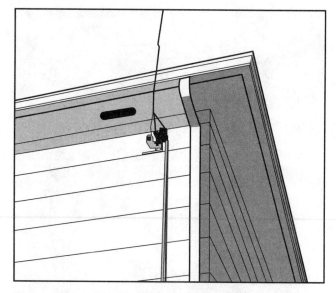

Fig. 3-12 *The other ends of the wires from the transformer are brought to the house and anchored there.*

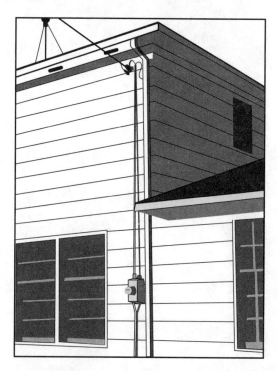

Fig. 3-13 *Wires from the pole are spliced to a larger three-wire Romex cable to carry power to the meter and then down to the distribution panel in the basement.*

simple and ancient: the wheel, the lever, the block and tackle, and the plow. For most tasks, farmers could draw on only their own strength, or that of horses and other animals. Children studied by the dim light of a kerosene lamp. Women were slaves to the wood stove and washboard.

Federal Agencies

In the United States, the federal government does not provide retail electric service over distribution systems, to either urban or rural areas. The role of government in the electric power industry is fourfold:

1. The Federal Power Commission exercises licensing authority over the utilization of hydroelectric sites on navigable rivers of the country. It also maintains certain controls over the interstate transmission and sale at wholesale of electric energy generated by electric companies.

2. The Bureau of Reclamation, the Army Corps of Engineers, and the Tennessee Valley Authority (TVA) build and operate some of the nation's hydroelectric generating plants. The TVA also builds steam plants.

3. The Bureau of Reclamation, TVA, the Bonneville Power Administration, and similar agencies build and operate transmission lines for marketing power wholesale.

4. The Rural Electrification Administration (REA) makes loans for rural electrification, including generation, transmission, and distribution systems.

Electricity for Everyone

The notion that electricity generated at a central station could be distributed to every farm in the United States took hold of people's minds slowly. Electric service was theoretically within the reach of rural families when the discovery was made that alternating-current voltage can be "stepped up" for transmission and "stepped down" for utilization. This, of course, means that power can be delivered economically to areas that are distant from the generating station.

Of course, technological theory was not sufficient to bring electrical power to rural areas. Financing on a large scale had to be provided. In the United States, farmers live on the land they cultivate. Farmhouses, therefore, are widely scattered across the countryside. In some ranch areas of the western states, houses are many miles apart. In such thinly populated rural areas, electric companies could see little prospect for profit. Farmers usually were required to pay construction costs of individual line extensions to provide service. Rates were high. Most could not afford this and, therefore, remained without electricity.

REA

In 1935, President Franklin D. Roosevelt created the Rural Electrification Administration as an emergency relief program. In May 1936, Congress passed the Rural Electrification Act. This established REA as a lending agency of the federal government. It had the responsibility of developing a program for rural electrification. The act authorized and empowered REA to make self-liquidating loans to companies, cooperatives, municipalities, and public power districts. These loans were to finance the construction and operation of generating plants, transmission and distribution lines, and related facilities for the purpose of furnishing electricity to persons in rural areas (Fig. 3-16).

In addition to making loans, REA furnishes technical assistance to its borrowers. It offers advice on engineering, management, accounting, public relations, power use, and legal matters. REA, however, does not construct, own, or operate electric facilities.

Rural distribution of power A typical rural distribution system serves a relatively large geographic area, usually with a uniform, light-load demand. An average system serves 5000 consumers with 1500 miles (2400

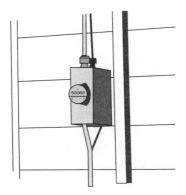

Fig. 3-14 *Close-up of a box containing a meter. The kilowatthour meter can be removed from the plug-in socket if necessary.*

Fig. 3-15 *This is a five-in-line clock dial, single-phase, 240-volt, three-wire kilowatthour meter.*

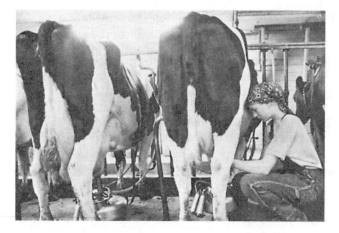

Fig. 3-16 *Dairy farmers use electric cow milkers.*

kilometers) of distribution line. However, some rural systems serve more than 25,000 customers. Here, farms are quite large, generally averaging 300 acres (121 hectares) or more. Typically, about three farms are served by each mile of distribution line. Loads are predominately single-phase, although a few three-phase installations may be located within the area. Towns, villages, and cities are served by separate distribution facilities not financed by REA. The typical rural system load is growing rapidly. It requires good voltage regulation and a high level of service continuity (Fig. 3-17).

These factors have a major influence on design and construction of a rural distribution system. The preponderance of single-phase loads has led to systems that are essentially single-phase in construction and operation. Approximately 80% of the lines being used are single-phase. The system that has evolved for rural service is typically an overhead multigrounded, neutral radial system on wood poles. It uses relatively

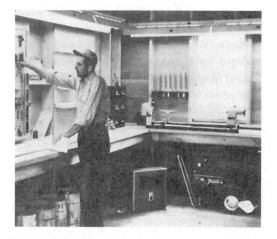

Fig. 3-17 *Today's farmer often has use of a complete workshop due to the availability of inexpensive electrical power.*

small conductors that are usually aluminum conductors with steel cores (ACSR).

The most common voltage is 7200 volts phase-to-ground, or 12,470 volts phase-to-phase. There are some sparsely settled areas remote from power sources where the individual farm loads are heavy. For such areas, it has been more economical to use 14.4/24.9 kilovolts as a distribution voltage.

Compared to other types of systems, the multi-grounded neutral system offers a number of distinct advantages for serving the remote area just described. Long-span, single-phase branches may be constructed most economically in vertical configuration. This is the basic line-supporting structure used in U.S. rural electrification.

Telephone and power lines The experience of rural distribution systems operating in close proximity to telephone or other communication systems has been satisfactory. The two systems frequently parallel each other at roadside separation. In addition, a large amount of line is built with the telephone and electric systems sharing the same poles. No special construction is required for the electric distribution system. However, the transformers used are built to a specification that controls the harmonic content of the excitation current. Telephone systems in the United States have determined that the most satisfactory method of operating is to adopt construction standards that are characterized by a very low susceptibility to power line influence. These standards make it unnecessary for telephone companies to request special construction from power systems to reduce power line influence on their communication cables.

Single-phase power The principal reason for favoring single-phase rural distribution lines is that most rural loads are single-phase. An average farm has no three-phase requirements. Most farm motors, including those used for feed grinding, are below 5 horsepower. Larger and deeper irrigation wells, however, frequently require pump motors of 25 horsepower or more, for which three-phase service is provided. In the past, 7½ horsepower generally was regarded as the largest practical single-phase motor. The cost of this motor was about double that of a comparable three-phase motor (Fig. 3-18).

Several hundred dollars for an occasional motor seems preferable to thousands of dollars for three-phase lines. Meanwhile, manufacturers are becoming interested in larger single-phase motors that will perform as well as, yet cost no more than, three-phase motors. Also note that many rural systems have the

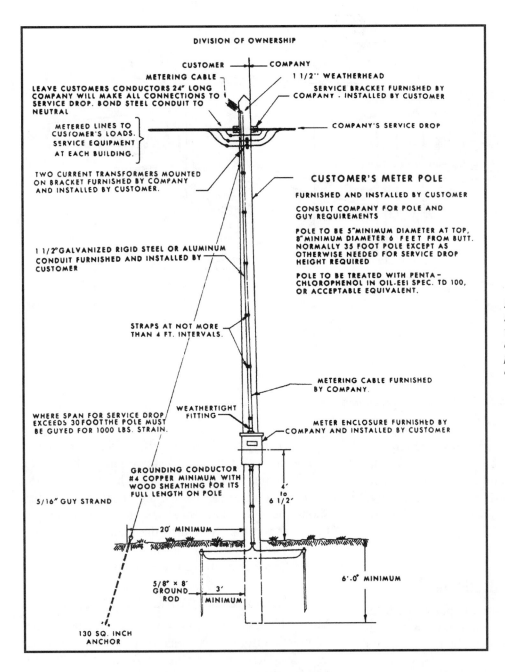

DIVISION OF OWNERSHIP

CUSTOMER — COMPANY

CUSTOMER
METERING CABLE — 1 1/2" WEATHERHEAD

LEAVE CUSTOMERS CONDUCTORS 24" LONG
COMPANY WILL MAKE ALL CONNECTIONS TO
SERVICE DROP. BOND STEEL CONDUIT TO
NEUTRAL

SERVICE BRACKET FURNISHED BY
COMPANY - INSTALLED BY CUSTOMER

COMPANY'S SERVICE DROP

METERED LINES TO
CUSTOMER'S LOADS.
SERVICE EQUIPMENT
AT EACH BUILDING.

TWO CURRENT TRANSFORMERS MOUNTED
ON BRACKET FURNISHED BY COMPANY
AND INSTALLED BY CUSTOMER.

CUSTOMER'S METER POLE

FURNISHED AND INSTALLED BY CUSTOMER

CONSULT COMPANY FOR POLE AND
GUY REQUIREMENTS

POLE TO BE 5" MINIMUM DIAMETER AT TOP,
8" MINIMUM DIAMETER 6 FEET FROM BUTT.
NORMALLY 35 FOOT POLE EXCEPT AS
OTHERWISE NEEDED FOR SERVICE DROP
HEIGHT REQUIRED

1 1/2" GALVANIZED RIGID STEEL OR ALUMINUM
CONDUIT FURNISHED AND INSTALLED BY
CUSTOMER

POLE TO BE TREATED WITH PENTA-
CHLOROPHENOL IN OIL-EEI SPEC. TD 100,
OR ACCEPTABLE EQUIVALENT.

STRAPS AT NOT MORE
THAN 4 FT. INTERVALS.

METERING CABLE FURNISHED
BY COMPANY.

WHERE SPAN FOR SERVICE DROP
EXCEEDS 30 FOOT THE POLE MUST
BE GUYED FOR 1000 LBS. STRAIN.

WEATHERTIGHT
FITTING

METER ENCLOSURE FURNISHED BY
COMPANY AND INSTALLED BY CUSTOMER

GROUNDING CONDUCTOR
#4 COPPER MINIMUM WITH
WOOD SHEATHING FOR ITS
FULL LENGTH ON POLE

4'
to
6 1/2'

5/16" GUY STRAND

20' MINIMUM

5/8" x 8'
GROUND
ROD

3'
MINIMUM

6'-0" MINIMUM

130 SQ. INCH
ANCHOR

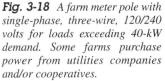

Fig. 3-18 *A farm meter pole with single-phase, three-wire, 120/240 volts for loads exceeding 40-kW demand. Some farms purchase power from utilities companies and/or cooperatives.*

capacity to serve 30-horsepower, single-phase motors without undue flicker problems. In fact, single-phase loads no longer are limited to any specific size. The present practice is to allow the capability of the system, at the time of installation, to determine the maximum size of the single-phase load that can be served.

INSTALLATION OF SERVICE BY POWER COMPANY

You have, no doubt, noticed that there are some standards for the transmission of electrical power over distances. The towers, poles, and supports of various types are standardized as to size and treatment to en-

dure years of exposure. There are some very definite minima and maxima for the crossing of roads and various other obstacles on the way from the generator to the user. Figure 3-19 shows the minimum clearance of service drops below 600 volts.

Figure 3-20 shows the attention given to the service entrance riser support on a low building. This work is usually done by the electrician, and at least 24-inch service conductors are left for connection by the power company when it brings power up to the house or building.

Most new suburbs are demanding that electrical service be brought in from the line to the house by

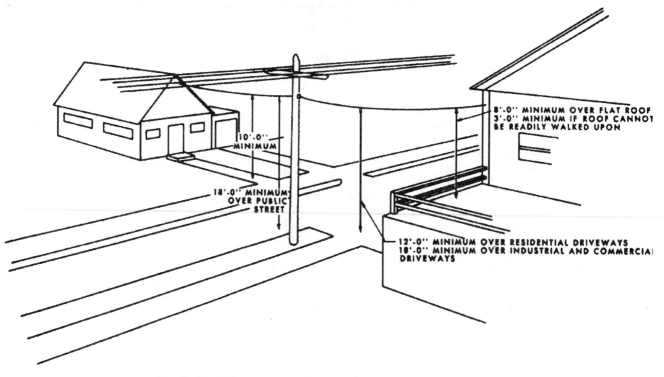

Fig. 3-19 *Minimum vertical clearance of service drops below 600 volts.*

underground cables. Figure 3-21 shows the requirements for such an installation. Note how far the meter is mounted above ground. This service shows the pole in the rear of the house. Some localities also require that the entire electrical service to an area be underground. This eliminates poles in the rear of the house.

Meter Installations

The electrician is usually required to place the meter socket trough. It is usually supplied by the power company and installed by the customer. Figure 3-22 shows a typical meter installation with a socket for plugging in the meter once the service has been connected. In the case of apartment houses, it will be necessary in some cases to install two or more meters.

Note in Fig. 3-23 (page 49) that six more meters can be installed indoors in a single-phase, three-wire, 120/240-volt service or single-phase, three-wire, 120/208-volt service. Notice the distance the meters should be above the floor. These are the plug-in type of meter sockets. Note also the materials furnished by the power company and installed by the electrician. The meter board is usually painted black.

Mobile Home Installation

Mobile homes have been getting an increasing amount of attention from the National Electric Code and other agencies. They are becoming a *permanent* dwelling for many people. These homes have particular needs that must be met to keep them electrically safe. Figure 3-24 shows a typical installation for mobile home parks.

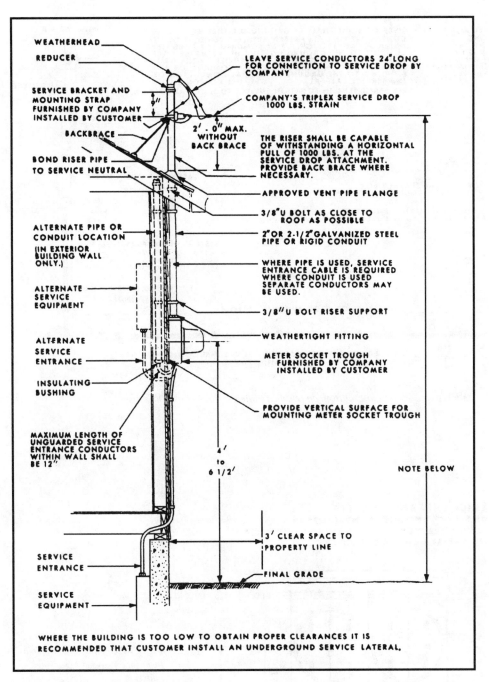

WEATHERHEAD

REDUCER

SERVICE BRACKET AND
MOUNTING STRAP
FURNISHED BY COMPANY
INSTALLED BY CUSTOMER

BACKBRACE

BOND RISER PIPE
TO SERVICE NEUTRAL

ALTERNATE PIPE OR
CONDUIT LOCATION
(IN EXTERIOR
BUILDING WALL
ONLY.)

ALTERNATE
SERVICE
EQUIPMENT

ALTERNATE
SERVICE
ENTRANCE

INSULATING
BUSHING

MAXIMUM LENGTH OF
UNGUARDED SERVICE
ENTRANCE CONDUCTORS
WITHIN WALL SHALL
BE 12"

SERVICE
ENTRANCE

SERVICE
EQUIPMENT

LEAVE SERVICE CONDUCTORS 24"LONG
FOR CONNECTION TO SERVICE DROP BY
COMPANY

COMPANY'S TRIPLEX SERVICE DROP
1000 LBS. STRAIN

9"

2' - 0" MAX.
WITHOUT
BACK BRACE

THE RISER SHALL BE CAPABLE
OF WITHSTANDING A HORIZONTAL
PULL OF 1000 LBS. AT THE
SERVICE DROP ATTACHMENT.
PROVIDE BACK BRACE WHERE
NECESSARY.

APPROVED VENT PIPE FLANGE

3/8"U BOLT AS CLOSE TO
ROOF AS POSSIBLE

2" OR 2-1/2"GALVANIZED STEEL
PIPE OR RIGID CONDUIT

WHERE PIPE IS USED, SERVICE
ENTRANCE CABLE IS REQUIRED
WHERE CONDUIT IS USED
SEPARATE CONDUCTORS MAY
BE USED.

3/8"U BOLT RISER SUPPORT

WEATHERTIGHT FITTING

METER SOCKET TROUGH
FURNISHED BY COMPANY
INSTALLED BY CUSTOMER

PROVIDE VERTICAL SURFACE FOR
MOUNTING METER SOCKET TROUGH

4'
to
6 1/2'

NOTE BELOW

3' CLEAR SPACE TO
PROPERTY LINE

FINAL GRADE

WHERE THE BUILDING IS TOO LOW TO OBTAIN PROPER CLEARANCES IT IS
RECOMMENDED THAT CUSTOMER INSTALL AN UNDERGROUND SERVICE LATERAL.

Fig. 3-20 Service entrance riser
support on a low building.

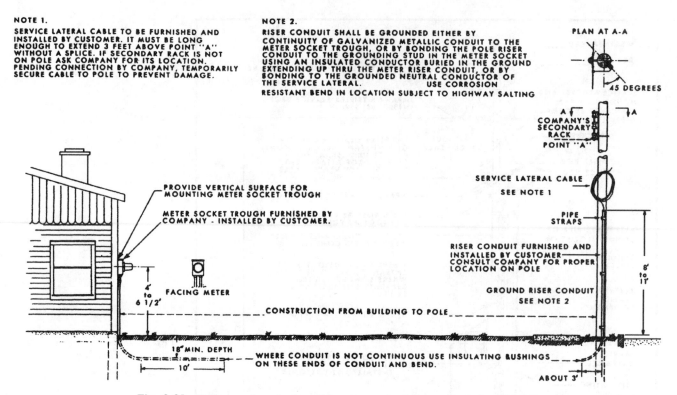

NOTE 1.

SERVICE LATERAL CABLE TO BE FURNISHED AND INSTALLED BY CUSTOMER. IT MUST BE LONG ENOUGH TO EXTEND 3 FEET ABOVE POINT "A" WITHOUT A SPLICE. IF SECONDARY RACK IS NOT ON POLE ASK COMPANY FOR ITS LOCATION. PENDING CONNECTION BY COMPANY, TEMPORARILY SECURE CABLE TO POLE TO PREVENT DAMAGE.

NOTE 2.

RISER CONDUIT SHALL BE GROUNDED EITHER BY CONTINUITY OF GALVANIZED METALLIC CONDUIT TO THE METER SOCKET TROUGH, OR BY BONDING THE POLE RISER CONDUIT TO THE GROUNDING STUD IN THE METER SOCKET USING AN INSULATED CONDUCTOR BURIED IN THE GROUND EXTENDING UP THRU THE METER RISER CONDUIT, OR BY BONDING TO THE GROUNDED NEUTRAL CONDUCTOR OF THE SERVICE LATERAL. USE CORROSION RESISTANT BEND IN LOCATION SUBJECT TO HIGHWAY SALTING

PLAN AT A-A

45 DEGREES

A
COMPANY'S SECONDARY RACK
POINT "A"

SERVICE LATERAL CABLE
SEE NOTE 1

PROVIDE VERTICAL SURFACE FOR MOUNTING METER SOCKET TROUGH

METER SOCKET TROUGH FURNISHED BY COMPANY - INSTALLED BY CUSTOMER.

PIPE STRAPS

RISER CONDUIT FURNISHED AND INSTALLED BY CUSTOMER CONSULT COMPANY FOR PROPER LOCATION ON POLE

GROUND RISER CONDUIT
SEE NOTE 2

FACING METER

4' to 6 1/2'

8' to 11'

CONSTRUCTION FROM BUILDING TO POLE

18" MIN. DEPTH

WHERE CONDUIT IS NOT CONTINUOUS USE INSULATING BUSHINGS ON THESE ENDS OF CONDUIT AND BEND.

10'

ABOUT 3'

Fig. 3-21 *Underground service lateral from an overhead line service below 600 volts.*

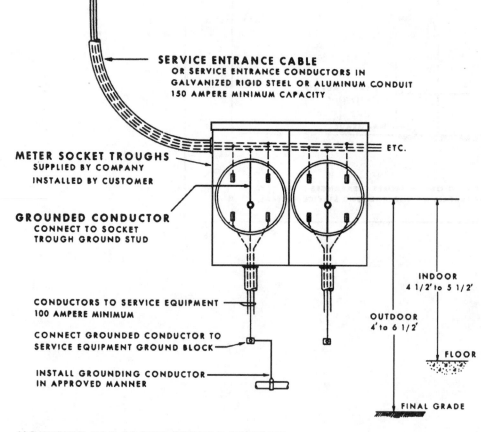

SERVICE ENTRANCE CABLE
OR SERVICE ENTRANCE CONDUCTORS IN GALVANIZED RIGID STEEL OR ALUMINUM CONDUIT 150 AMPERE MINIMUM CAPACITY

ETC.

METER SOCKET TROUGHS
SUPPLIED BY COMPANY
INSTALLED BY CUSTOMER

GROUNDED CONDUCTOR
CONNECT TO SOCKET TROUGH GROUND STUD

Fig. 3-22 *Two- to six-meter installation, socket type. Single-phase, three-wire, 120/240-volt, 150-ampere minimum service entrance.*

INDOOR
4 1/2' to 5 1/2'

OUTDOOR
4' to 6 1/2'

CONDUCTORS TO SERVICE EQUIPMENT 100 AMPERE MINIMUM

CONNECT GROUNDED CONDUCTOR TO SERVICE EQUIPMENT GROUND BLOCK

INSTALL GROUNDING CONDUCTOR IN APPROVED MANNER

FLOOR

FINAL GRADE

MOUNTING FOR SOCKET TROUGH REQUIRES:
OUTDOORS - A VERTICAL MOUNTING SURFACE.
INDOORS - A METER BOARD OF 3/4" PLYWOOD, PAINTED AND SECURELY MOUNTED.

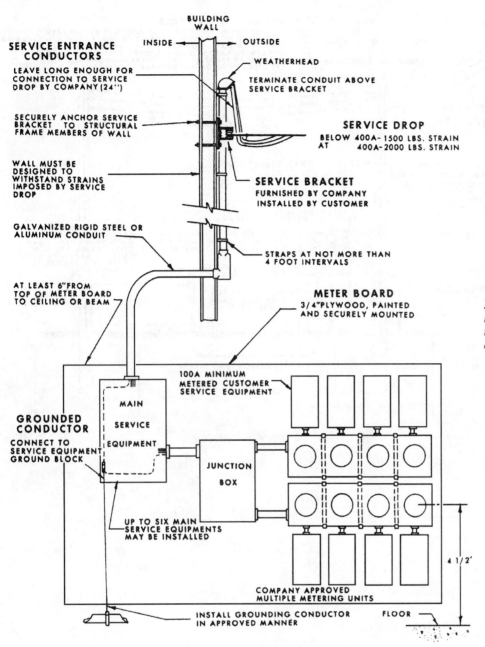

SERVICE ENTRANCE CONDUCTORS

LEAVE LONG ENOUGH FOR CONNECTION TO SERVICE DROP BY COMPANY (24")

SECURELY ANCHOR SERVICE BRACKET TO STRUCTURAL FRAME MEMBERS OF WALL

WALL MUST BE DESIGNED TO WITHSTAND STRAINS IMPOSED BY SERVICE DROP

GALVANIZED RIGID STEEL OR ALUMINUM CONDUIT

AT LEAST 6" FROM TOP OF METER BOARD TO CEILING OR BEAM

GROUNDED CONDUCTOR

CONNECT TO SERVICE EQUIPMENT GROUND BLOCK

UP TO SIX MAIN SERVICE EQUIPMENTS MAY BE INSTALLED

BUILDING WALL

INSIDE   OUTSIDE

WEATHERHEAD

TERMINATE CONDUIT ABOVE SERVICE BRACKET

SERVICE DROP

BELOW 400A - 1500 LBS. STRAIN
AT   400A - 2000 LBS. STRAIN

SERVICE BRACKET

FURNISHED BY COMPANY INSTALLED BY CUSTOMER

STRAPS AT NOT MORE THAN 4 FOOT INTERVALS

METER BOARD

3/4" PLYWOOD, PAINTED AND SECURELY MOUNTED

100A MINIMUM METERED CUSTOMER SERVICE EQUIPMENT

MAIN SERVICE EQUIPMENT

JUNCTION BOX

COMPANY APPROVED MULTIPLE METERING UNITS

INSTALL GROUNDING CONDUCTOR IN APPROVED MANNER

FLOOR

4 1/2'

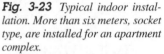

Fig. 3-23 *Typical indoor installation. More than six meters, socket type, are installed for an apartment complex.*

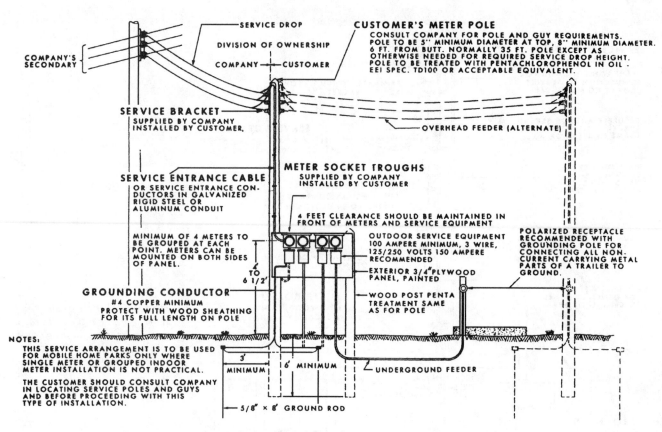

Fig. 3-24 *A typical installation for a mobile home park.*

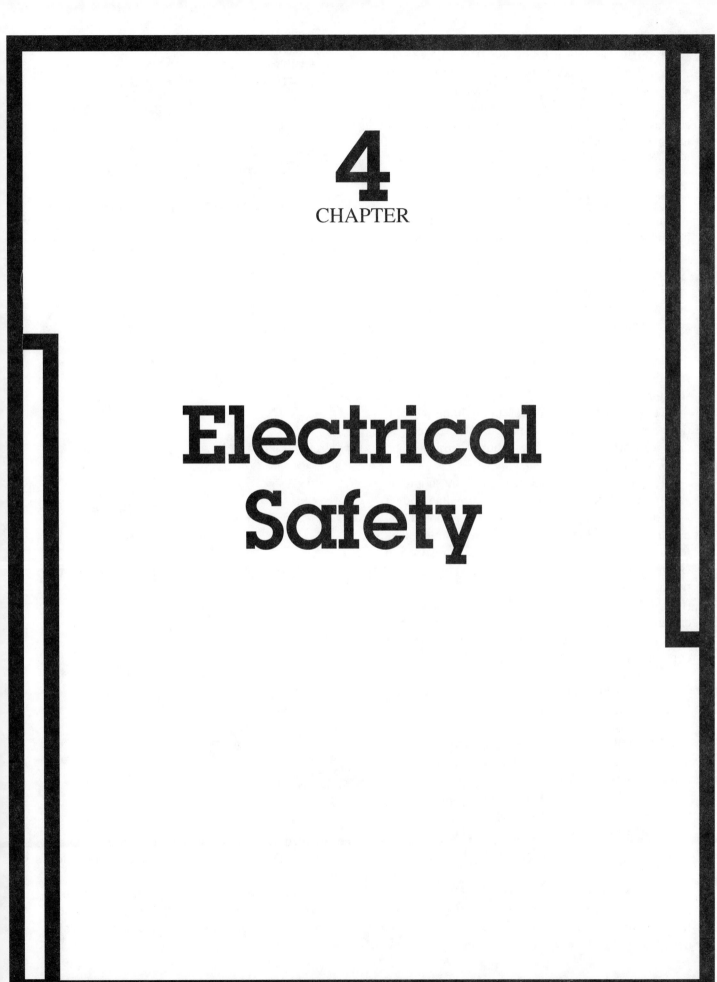

4
CHAPTER

Electrical Safety

SAFETY ON THE JOB IS VERY IMPORTANT FOR the electrician. A number of safety procedures are needed to make certain that everyone on a construction site is protected. A building under construction can be an unsafe place. Thus, it is important that certain minimum conditions be met in terms of electrical wiring.

TEMPORARY WIRING

A temporary electrical wiring system does not have to be made with the detail and relative permanence needed for permanent wiring systems. However, the National Electrical Code (NEC) deals with this type of installation in Article 305. The article deals with the only ways a temporary wiring system may differ from a permanent system.

The NEC says that temporary electrical power and lighting installations shall be permitted during the construction, remodeling, or demolition of buildings, structures, or equipment.

Temporary electrical power and lighting installations are permitted during a period not to exceed 90 days for Christmas decorative lighting, carnivals, and experimental or development work.

Temporary feeders may be used. They should be supported by insulators spaced not more than 10 feet apart if they are open conductors. They should terminate in a suitable or approved power outlet. This means one that is made by the manufacturer for job site temporary wiring. Several variations of protection may be provided by portable receptacle boxes (Fig. 4-1). These portable units have ground fault circuit interrupters (GFCIs) located within them. They protect the entire circuit. A number of different cord sets are available for use with GFCI-protected plug-in units.

Most residential builders use a nonmetallic feeder and run it to a nonprotected receptacle. This does not provide adequate protection for the power tool user. Other parts of this chapter will show how the short circuits in a power tool can cause serious shock hazards.

A responsible type of protection is preferred. It will pay for itself in a few jobs inasmuch as it is built to last. It does not have to be "jerry-rigged" each time a new construction job is undertaken.

TEMPORARY LIGHTING

Article 305-4 (Part f) of the NEC states that all lamps for general illumination shall be protected from accidental contact or breakage (Fig. 4-2). Protection shall

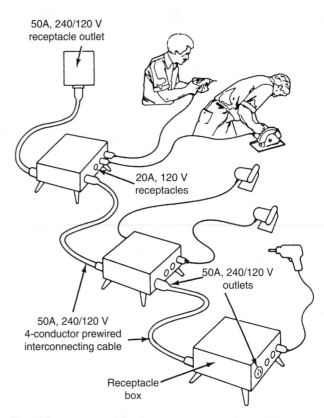

Fig. 4-1 *Temporary branch circuits are manufactured for use at construction sites.*

50A, 240/120 V receptacle outlet

20A, 120 V receptacles

50A, 240/120 V outlets

50A, 240/120 V 4-conductor prewired interconnecting cable

Receptacle box

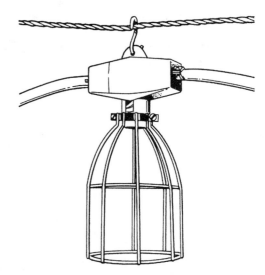

Fig. 4-2 *Temporary lighting strings of cable and sockets are available manufactured.*

be provided by elevation of at least 7 feet from normal working surface or by a suitable fixture or lampholder with a guard.

Figure 4-3 shows how temporary wiring may be installed for use at a construction site. Splices may be made without boxes for the cord and cable runs.

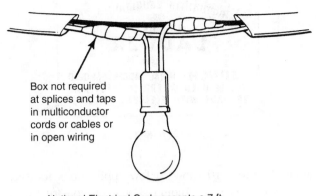

Temporary wiring for new construction

Box not required at splices and taps in multiconductor cords or cables or in open wiring

National Electrical Code accepts a 7 ft mounting height as adequate protection without a guard.

Fig. 4-3 *Splices made without boxes for temporary wiring of lights at a construction site.*

TEMPORARY CORDS

Much information is available on extension cords that can be used on a construction site. More is presented later in the chapter. Special watertight plugs and connectors provide insurance against nuisance tripping of the GFCI (Fig. 4-4).

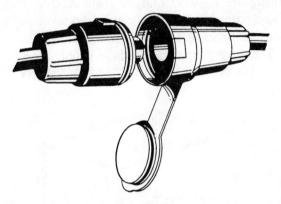

Fig. 4-4 *Special watertight plugs and connectors are suggested for use at construction sites.*

LADDERS

It is necessary for an electrician to work with outlets and fixtures that are above his or her reach. This requires a ladder. There are three types of ladders: folding, straight, and extension (Fig. 4-5).

Metal and water conduct electricity. Do not use a metal, metal-reinforced, or wet ladder where direct contact is possible with a power source.

For safety on a ladder, the pitch or angle should be such that the horizontal distance at the bottom is one-fourth the working length of the ladder (Fig. 4-6). Keep in mind that the extension ladder should be

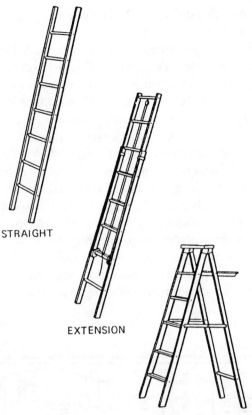

STRAIGHT

EXTENSION

FOLDING

Fig. 4-5 *Three types of ladders.*

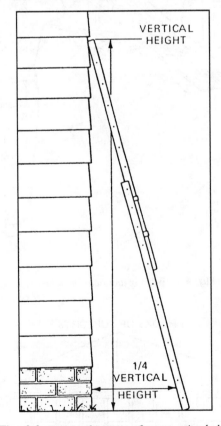

VERTICAL HEIGHT

1/4 VERTICAL HEIGHT

Fig. 4-6 *Proper placement of an extension ladder.*

placed so that the top of the ladder extends above the edge of the roof at least 3 feet.

To raise an extension-type ladder, walk forward under the ladder, as shown in Fig. 4-7. Move your hands to grasp other rungs as you proceed forward.

When you have finished, fold the ladder. Place it where it is not exposed to the weather or excess heat. To prevent sagging, an extension ladder should be stored horizontally on supports (Fig. 4-8).

Underwriters Laboratories (UL) tests ladders. A label bearing its seal is placed on tested ladders (Fig. 4-9). This seals warns that metal ladders should not be used where contact may be made with electrical circuits. Some electrical inspectors require that all metal ladders used on a job site have this seal.

In some residential construction, high ceilings make it necessary to construct scaffolding to reach some work

Fig. 4-9 *The UL seal for a ladder.*

areas. Figure 4-10 shows a few applications for such scaffolding. The scaffolding should be constructed so that it is absolutely secure. Makeshift scaffolding has been the cause of many accidents. It should be maneuverable and capable of being locked in position. Remove the scaffolding as soon as its use has been served. Store it in a safe location so it will not block any passageway.

There are a number of devices designed to protect people from their carelessness and from electrical shock. If you work around electricity for long, you will receive a shock; most likely it will be caused by your own negligence. Such a shock could be fatal. There is very little difference between the experience of a slight tingle and a fatal shock. Eight milliamperes is 0.008 ampere, and 50 milliamperes is 0.050 ampere. There is, then, only a 0.042-ampere difference between the shock you feel as a tingle and the amount of current that will *kill* you.

There have been cases where people have been shocked or burned by high voltages. Voltage does not

Fig. 4-7 *Walking an extension ladder up.*

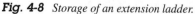

HOOKS OR SUPPORTS

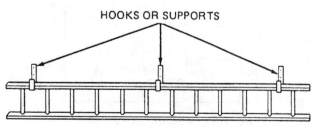

Fig. 4-8 *Storage of an extension ladder.*

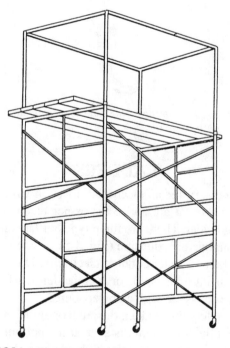

Fig. 4-10A *Adjustable scaffolding may be used by an electrician on the job.*

Fig. 4-10B *Manufactured scaffolding may be made with a platform to support the electrician.*

necessarily kill—it is the *amount of current* through the body that does the damage.

These facts and figures are presented, not to scare you, but to make you aware of the dangers of being careless with the power in wires. This chapter deals with some of the aspects of safety in electrical work. There are ways to be safe and live a long life. There are careless practices that are bound to mean a shorter life. Read this safety section with the idea that the life saved could easily be your own.

FATAL CURRENT

Strange as it seems, most fatal electric shocks happen to people who should know better. Here are some "electromedical" facts that should make you think twice before taking chances.

It's the Current That Kills

Offhand, it would seem that a shock of 10,000 volts would be more deadly than a shock of 100 volts. But this is not so. Individuals have been electrocuted by appliances using ordinary house current of 110 volts and by electrical apparatus in industry using as little as 42 volts direct current (DC). The real measure of a shock's intensity lies in the amount of current (milliamperes) forced through the body, and not in the voltage. Any electrical device used on a house wiring circuit can, under certain conditions, transmit a fatal current.

Any amount of current over 10 milliamperes (0.01 ampere) is capable of producing painful to severe shock. However, currents between 100 and 200 mA (0.1 to 0.2 ampere) are lethal.

From a practical viewpoint, after a person is knocked out by electric shock, it is impossible to tell how much current passed through the vital organs of the body. Artificial respiration must be applied immediately if breathing has stopped.

Physiological effects of electric shock   Table 4-1 shows the physiological effects of various current densities. Note that voltage is not a consideration. It takes voltage to make current flow. However, the amount of shock current will vary, depending on the body resistance between the points of contact.

As shown in Table 4-1, shock is more severe as the current rises. At values as low as 20 mA, breathing becomes labored, finally ceasing completely even at values below 75 mA.

As current approaches 100 mA, ventricular fibrillation of the heart occurs—an uncoordinated twitching of the walls of the heart's ventricles.

NOTE: A heart that is in fibrillation cannot be restricted by closed-chest cardiac massage. A special device called a *defibrillator* is available in some medical facilities and by ambulance services.

With more than 200 mA, muscular contractions are so severe that the heart is forcibly clamped during the shock. This clamping prevents the heart from going into ventricular fibrillation, making the victim's chances for survival better.

Table 4-1 *Physiological Effects of Electric Currents**

Readings		Effects
Safe current values	1 mA or less	Causes no sensation—not felt.
	1 to 8 mA	Sensation of shock, not painful; individual can let go at will since muscular control is not lost.
Unsafe current values	8 to 15 mA	Painful shock; individual can let go at will since muscular control is not lost
	15 to 20 mA	Painful shock; control of adjacent muscles is lost; victim cannot let go.
	20 to 50 mA	Painful, severe muscular contractions; breathing is difficult.
	50 to 100 mA	Ventricular fibrillation—a heart condition that can result in instant death—is *possible*.
	100 to 200 mA	Ventricular fibrillation occurs.
	200+ mA	Severe burns, severe muscular contractions—so severe that chest muscles clamp the heart and stop it for the duration of the shock. (This prevents ventricular fibrillation.)

*Information provided by National Safety Council.

Danger—low voltage   It is common knowledge that victims of high-voltage shock usually respond to artificial respiration more readily than the victims of low-voltage shock. The reason may be the above-mentioned clamping of the heart, due to the high current densities associated with high voltages. However, to prevent a misinterpretation of those details, remember that 75 volts is just as lethal as 750 volts.

The actual resistance of the body varies, depending upon the points of contact and whether the skin is moist or dry. The area from one ear to the other, for example, has an internal resistance (which is lower than skin resistance) of only 100 ohms; from hand to foot it is nearer 500 ohms. Skin resistance may vary from 1000 ohms for wet skin, to more than 500,000 ohms for dry skin.

GENERAL SAFETY PRECAUTIONS

- *When you are working around electrical equipment, move slowly.* Make sure your feet are firmly placed for good balance. Don't lunge after falling tools. Kill all power and ground all high-voltage points before touching wiring. Make sure that power cannot be accidentally restored. Do not work on ungrounded equipment.

- *Do not examine live equipment when you are mentally or physically fatigued.* Keep one hand in your pocket while investigating live electrical equipment. Most important, do not touch electrical equipment while standing on metal floors, damp concrete, or other well-grounded surfaces. Do not handle electrical equipment while wearing damp clothing (particularly wet shoes) or while skin surfaces are damp.

- *Do not work alone!* Remember, the more you know about electrical equipment, the more heedless you're apt to become. Do not take unnecessary risks.

WHAT TO DO FOR VICTIMS

- *Cut voltage and/or remove victim from contact as quickly as possible—but without endangering your own safety.* Use a length of dry wood, rope, blanket, or similar device to pry or pull the victim loose. Don't waste valuable time looking for the power switch. The resistance of the victim's contact decreases with the passage of time. The fatal 100- to 200-mA level may be reached if action is delayed.

- *Start artificial respiration at once if the victim is unconscious and has stopped breathing.* Do not stop resuscitation until a medical authority pronounces the victim beyond help. It may take as long as 8 hours to revive a patient. There may be no pulse. A condition similar to rigor mortis may be present. However, these are the manifestations of shock, not necessarily indications that the victim has died.

Table 4-2 gives some of the resistances the human body presents to an applied voltage. If you want to know the amount of current that would pass through the body, use Ohm's law formulas and the applied voltage. Will it be enough to produce a fatal shock?

Table 4-2 *Human Resistance to Electric Current*

Body Area	Resistance (Ohms)
Dry skin	100,000 to 600,000
Wet skin	1000
Internal body—hand to foot	400 to 600
Ear to ear	(about) 100

$$\text{Current through body} = \frac{\text{voltage applied to body}}{\text{resistance of body and contacts}}$$

Shock severity depends upon the following factors:

- Becomes more severe with increased voltage
- Increases with the amount of moisture on contact surfaces
- Increases with an increase in pressure of contact
- Increases with an increase in area of body contact
- Resistance of body portions

The following three factors are involved in electric shock:

- Voltage (analogous to pressure forcing water in pipe)
- Current (analogous to water flowing in pipe)
- Resistance (analogous to something in pipe tending to hold back water flow)

Practical Problem

A worker with wet clothes, or wet with perspiration, comes in contact with a defective 120-volt light cord and establishes a good ground.

From Table 4-2,

Wet-skin resistance	1000 ohms
Body resistance	500 ohms
Internal resistance	1500 ohms

From the formula:

$$\text{Current through body} = \frac{120 \text{ volts}}{1500 \text{ ohms}} = 80 \text{ mA}$$

Note that muscular control is lost and the victim cannot break contact. As the contact is continued, the skin resistance is reduced. If the skin is punctured, skin resistance may then be disregarded. Then for practical purposes, the total resistance of the worker may be in the neighborhood of 600 ohms.

$$\text{Current through body} = \frac{120 \text{ volts}}{600 \text{ ohms}} = \frac{200 \text{ mA}}{\text{(certain death)}}$$

There is always a danger of an electrical wire with high voltages leaking to ground through another path (if the ground wire is broken). You may become the path of least resistance if you come too close (Table 4-3).

Table 4-3 Safe Distances from Live Circuits in Air

Do not approach live conductors closer than the following distances:	
751 to 3,500 volts	1 foot
3,501 to 10,000 volts	2 feet
10,001 to 50,000 volts	3 feet
50,001 to 100,000 volts	5 feet
100,001 to 250,000 volts	10 feet

To be safe, keep away from high-voltage wires. However, if you work with them or on them, make sure you use the proper safety procedures and clothing.

TYPICAL SHOCK HAZARDS WITH EXTENSION CORDS, PLUGS, AND RECEPTACLES
Extension Cords

A proper extension cord can make a difference in the safe operation of a piece of equipment. Make sure it has the proper capacity to handle the current needed. See Table 4-4.

Wire size and insulation of extension cords must be considered to ensure proper operation with a piece of equipment. Take a look at Table 4-5 for an explanation of the letters on a cord body. This table will help you determine the type of service for which a specific cord is recommended.

Length of an extension cord is important in the sense that it must have the correct size wire to allow full voltage to reach the consuming device. For instance, a 25-foot cord with No. 18 wire size is good for 2 amperes. If the distance is increased to 50 feet, you are still safe with No. 18 and 2 amperes. However, for a distance of 200 feet, the size of the wire must be increased to No. 16. This is necessary to carry the 2

Table 4-4 Flexible Cord Ampacities*

Size AWG	Type S, SO, ST	Type SJ, SJO, SJT
	Amperes	
18	7, 10†	7, 10†
16	10, 13†	10, 13†
14	15, 18†	
12	20	
10	25	
8	35	
6	45	
4	60	

*Table 400.5. Reprinted by permission from NFPA, National Electric Code, © 2002, National Fire Protection Association, Quincy, Massachusetts.
†Where third conductor is used for equipment grounding only and does not carry load current.

amperes without dropping the voltage along the cord and therefore producing a low voltage at the consuming device at the end of the line. See Table 4-6.

Inspection and preventive maintenance All drop or extension cords and other electrical equipment should be the proper size for the purpose intended. Wire size and insulation must be adequate for the service to be expected. Only equipment and cords listed by Underwriters Laboratories for the intended purpose should be used. Cords that are perfectly safe when purchased may be dangerous a short time later, unless they are given careful use. They should be kept free of oil and water. They should be protected from abrasion or other mechanical damage.

A system of preventive maintenance is suggested as a means of control over portable electrical equipment. Preventive maintenance involves inspection at regular intervals, the keeping of records of inspections, and making repairs when necessary. Splicing of extension cords should never be permitted.

A designated person, usually the electrician, should check the extension cords before they are put into use each day. The cord should have a continuous grounding conductor maintained throughout and carried back to the grounding bus at the service panel.

A visual inspection of the cord and harness should be made before each day's use. Check for loose or missing screws in the end and for loose or broken blades in the plug end (Fig. 4-11).

Fig. 4-11 *Check the extension cord before it is placed in service each day.*

Table 4-5 *Types of Flexible Cords*

Trade Name	Type Letter	Size AWG	No. of Conductors	Insulation Braid on	Braid on each Conductor	Outer Covering	Use
Junior hard-service cord	SJ SJO SJT	18, 14	2, 3 or 4	Rubber Thermoplastic or rubber	None	Rubber Oil resist, compound Thermoplastic	Pendant or portable Damp places Hard usage
Hard-service cord	S SO ST	6 18, to 10 incl.	2 or more	Rubber Thermoplastic or rubber	None	Rubber Oil resist, compound Thermoplastic	Pendant or portable Damp places Hard usage

Table 4-6 *Extension Cord Sizes for Portable Electric Tools*

THIS TABLE FOR 115-VOLT TOOLS						
Full-Load Ampere Rating of Tool	0 to 2.0 A	2.1 to 3.4 A	3.5 to 5.0 A	5.1 to 7.0 A	7.1 to 12.0 A	12.1 to 16.0 A
Length of Cord	Wire Size (AWG)					
25 ft	18	18	18	16	14	14
50 ft	18	18	18	16	14	12
75 ft	18	18	16	14	12	10
100 ft	18	16	14	12	10	8
200 ft	16	14	12	10	8	6
300 ft	14	12	10	8	6	4
400 ft	12	10	8	6	4	4
500 ft	12	10	8	6	4	2
600 ft	10	8	6	4	2	2
800 ft	10	8	6	4	2	1
1000 ft	8	6	4	2	1	0

Note: If voltage is already low at the source (outlet), have voltage increased to standard, or use a much larger cable than listed in order to prevent any further loss in voltage.

Check the continuity of each conductor with an ohmmeter at least once a week to make sure there are no broken conductors. Check the cord when it is first used or taken from the box. Check again after it has been repaired or after an incident such as shown in Fig. 4-12. If the plug has been repaired, check to see if the clamp on the wire is securely attached (Fig. 4-13).

Electric Plugs and Receptacles

Electric plugs and receptacles come in a wide variety of shapes and arrangements. Because there must be some regulation to ensure that plugs and receptacles are used properly, the National Electrical Manufacturers Association (NEMA) has developed a standard for manufacturing these devices. Table 4-7 shows how different

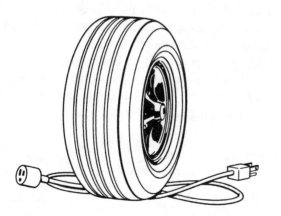

Fig. 4-12 *After an accident with construction equipment, the cord should be checked for damage.*

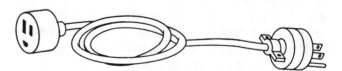

Fig. 4-13 *Check the wire clamp on the plug for proper fit.*

voltages and current combinations are specified for a particular plug or receptacle. Note the wiring diagram, shown for three-phase, single-phase, wye and delta connections, and wye and delta with various grounds. You have to be very careful when wiring these plugs to make sure the proper terminal is connected.

Do not use a 3φ plug or receptacle for 240-volt, three-wire, single-phase power.

Grounded conductor A *grounded conductor* has a white jacket in a two- or three-wire cable (neutral wire). It is terminated to the white, or silver-colored terminal in a plug cap or connector; and it is terminated at the neutral bar in the distribution box.

An electrical fault may allow the hot line to contact the metal housing of electrical equipment (in a typical two-wire system) or some other ungrounded

Table 4-7 *Voltage and Current Combinations for Plugs and Receptacles*

Wiring Diagram	NEMA ANSI	Receptacle Configuration	Rating		Rating	Receptacle Configuration	NEMA ANSI	Wiring Diagram
	5-15 C73.11		15A 125V	2 POLE 3 WIRE	30A 3φ 250V		11-30 C73.56		3P3W
	5-20 C73.12		20A 125V		50A 3φ 250V		11-50 C73.57	
	5-30 C73.45		30A 125V		15A 125/250V		14-15 C73.49		3 POLE 4 WIRE
	5-50 C73.46		50A 125V		20A 125/250V		14-20 C73.50	
	6-15 C37.20		15A 250V		30A 125/250V		14-30 C73.16	
	6-20 C73.51		20A 250V		50A 125/250V		14-50 C73.17	
	6-30 C73.52		30A 250V		60A 125/250V		14-60 C73.18	
	6-50 C73.53		50A 250V		15A 3φ 250V		15-15 C73.58	
	7-15 C73.28		15A 277V		20A 3φ 250V		15-20 C73.59	
	7-20 C73.63		20A 277V		30A 3φ 250V		15-30 C73.60	
	7-30 C73.64		30A 277V		50A 3φ 250V		15-50 C73.61	
	7-50 C73.65		50A 277V		60A 3φ 250V		15-60 C73.62	
	10-20 C73.23		20A 125/250V	3 POLE 3 WIRE	15A 3φY 120/208V		18-15 C73.15		4 POLE 4 WIRE
	10-30 C73.24		30A 125/250V		20A 3φY 120/208V		18-20 C73.26	
	10-50 C73.25		50A 125/250V		30A 3φY 120/208V		18-30 C73.47	
	11-15 C73.54		15A 3φ 250V		50A 3φY 120/208V		18-50 C73.48	
	11-20 C73.55		20A 3φ 250V		60A 3φY 120/208V		18-60 C73.27	

Table 4-7 *(Continued)*

	Rating	Receptacle Configuration	NEMA *ANSI*	Wiring Diagram
3P3W	30A 3φ 480V		L12-30 *C73.102*	
	30A 3φ 600V		L13-30 *C73.103*	
3 POLE 4 WIRE	20A 125/250V		L14-20 *C73.83*	
	30A 125/250V		L14-30 *C73.84*	
	20A 3φ 250V		L15-20 *C73.85*	
	30A 3φ 250V		L15-30 *C73.86*	
	20A 3φ 480V		L16-20 *C73.87*	
	30A 3φ 480V		L16-30 *C73.88*	
	30A 3φ 600V		L17-30 *C73.89*	
4 POLE 4 WIRE	20A 3φY 120/208V		L18-20 *C73.104*	
	30A 3φY 120/208V		L18-30 *C73.105*	
	20A 3φY 277/480V		L19-20 *C73.106*	
	30A 3φY 277/480V		L19-30 *C73.107*	
	20A 3φY 347/600V		L20-20 *C73.108*	
	30A 3φY 347/600V		L20-30 *C73.109*	
4P5W	20A 3φY 120/208V		L21-20 *C73.90*	
	30A 3φY 120/208V		L21-30 *C73.91*	
4 POLE 5 WIRE	20A 3φY 277/480V		L22-20 *C73.92*	
	30A 3φY 277/480V		L22-30 *C73.93*	
	20A 3φY 347/600V		L23-20 *C73.94*	
	30A 3φY 347/600V		L23-30 *C73.95*	

Wiring Diagram	NEMA *ANSI*	Receptacle Configuration	Rating	
	ML2 *C73.44*		15A 125V	**2 POLE 3 WIRE**
	L5-15 *C73.42*		15A 125V	
	L5-20 *C73.72*		20A 125V	
	L5-30 *C73.73*		30A 125V	
	L6-15 *C73.74*		15A 250V	
	L6-20 *C73.75*		20A 250V	
	L6-30 *C73.76*		30A 250V	
	L7-15 *C73.43*		15A 277V	
	L7-20 *C73.77*		20A 277V	
	L7-30 *C73.78*		30A 277V	
	L8-20 *C73.79*		20A 480V	
	L8-30 *C73.80*		30A 480V	
	L9-20 *C73.81*		20A 600V	
	L9-30 *C73.82*		30A 600V	
	ML3 *C73.30*		15A 125/250V	**3P3W**
	L10-20 *C73.96*		20A 125/250V	
	L10-30 *C37.97*		30A 125/250V	
	L11-15 *C73.98*		15A 3φ 250V	**3 POLE 3 WIRE**
	L11-20 *C73.99*		20A 3φ 250V	
	L11-30 *C73.100*		30A 3φ 250V	
	L12-20 *C73.101*		20A 3φ 480V	

Bryant Electric Co.

conductor. In these cases, any person who touches that equipment or conductor will receive a shock. The person completes the circuit from the hot line to the ground, and current passes through the body. Because a body is not a good conductor, the current is not high enough to blow the fuse. It continues to pass through the body as long as the body remains in contact with the equipment (Fig. 4-14).

Grounding conductor A *grounding conductor,* or equipment ground, is a wire attached to the housing, or other conductive parts of electrical equipment that are not normally energized, to carry current from it or them to the ground. Thus, if a person touches a part that is accidentally energized, there will be no shock, because the grounding line furnishes a much lower resistance path to the ground (Fig. 4-15). Moreover, the high current passing through the wire conductor blows the fuse and stops the current. In normal operation, a grounding conductor does not carry current.

The grounding conductor in a three-wire conductor cable has a green jacket. It is always terminated at the green hex-head screw on the cap or connector. It utilizes either a green conductor or a metallic conduit as its path to ground. In Canada, this conductor is referred to as the *earthing conductor.* This term is somewhat more descriptive and helpful in distinguishing between grounding conductors and neutral wires, or grounded conductors.

Two-wire configurations The standard parallel configuration, illustrated in Fig. 4-16, is designated a 125-volt, 15-ampere, two-pole, two-wire type at present. It will be phased out as old wiring is replaced with newer, two-wire with ground plugs and receptacles. These older types are no longer permitted by the National Electrical Code.

A polarized parallel configuration (Fig. 4-17) is also designated a 125-volt, 15-ampere, two-pole, two-wire type. It differs from the standard parallel only in that one of the blades is about $1/16$ inch wider than the other. This wider blade is intended for connection to the neutral wire. Cord connectors and receptacles in the configuration also have one wide slot. This ensures that,

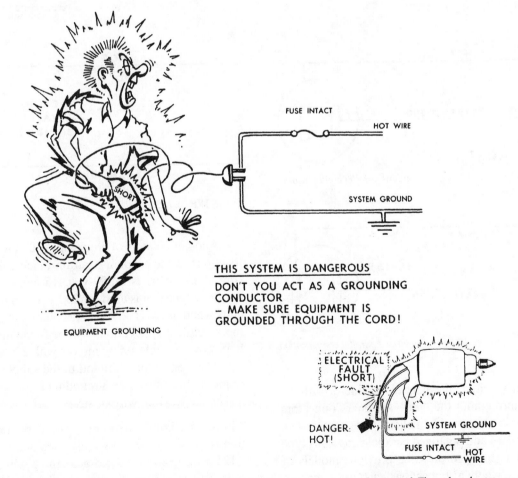

Fig. 4-14 *White-jacketed system grounds cannot connect electricity from short circuits to the ground. Thus, they do not prevent the housing of faulty equipment from being charged. Therefore, a person who contacts the charged housing becomes the conductor in a short circuit to the ground.*

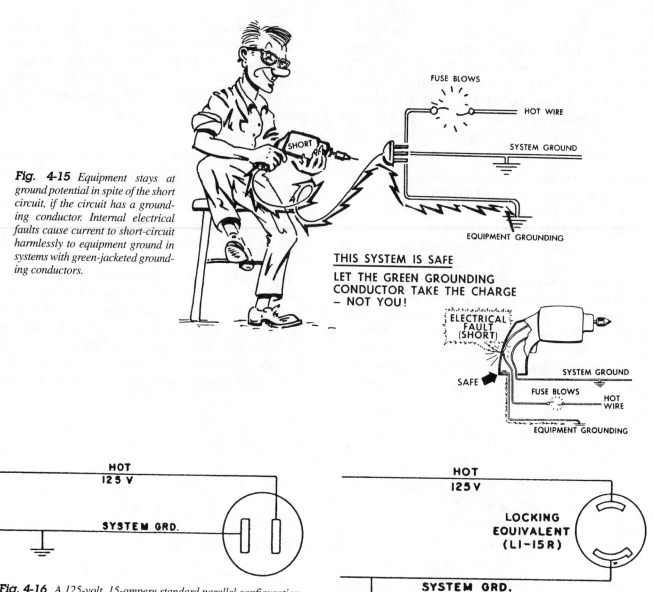

Fig. 4-15 *Equipment stays at ground potential in spite of the short circuit, if the circuit has a grounding conductor. Internal electrical faults cause current to short-circuit harmlessly to equipment ground in systems with green-jacketed grounding conductors.*

FUSE BLOWS

HOT WIRE

SYSTEM GROUND

EQUIPMENT GROUNDING

THIS SYSTEM IS SAFE

LET THE GREEN GROUNDING CONDUCTOR TAKE THE CHARGE — NOT YOU!

ELECTRICAL FAULT (SHORT)

SAFE

SYSTEM GROUND

FUSE BLOWS

HOT WIRE

EQUIPMENT GROUNDING

HOT
125 V

SYSTEM GRD.

Fig. 4-16 *A 125-volt, 15-ampere standard parallel configuration.*

HOT
125 V

(L-15R)

SYSTEM GRD.

Fig. 4-17A *A 125-volt, 15-ampere polarized parallel configuration.*

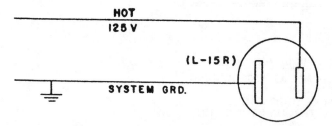

HOT
125 V

LOCKING EQUIVALENT (L1-15R)

SYSTEM GRD.

Fig. 4-17B *A 125-volt, 15-ampere polarized locking configuration.*

when properly wired, the neutral conductor will be continuous throughout the entire system. Connectors have one wide slot, enabling them to accommodate both the standard and polarized parallel caps. This type of plug may be noticed on some previous models of television sets, where a "hot chassis" was used, making it important to connect the ground wire to the chassis.

Another old type of plug (cap) you may see in some locations is the tandem-position blades, now banned (Fig. 4-18). This is a 250-volt, 15 ampere, two-pole, two-wire type that is used for 230-volt, single-phase circuits where there is no separate equipment ground. The most frequent applications will be found on fractional-horsepower, single-phase motors with 230 volts, where a separate equipment ground in the supply cord is not required. In some cases, grounding is accomplished by a separate flexible wire or strap.

Three-wire configurations The U-blade configuration (Fig. 4-19) is most widely used and is designated a 125-volt, 15-ampere, two-pole, three-wire, grounding type. The grounding blade on the plug is slightly longer than the two line blades. The configuration is used for

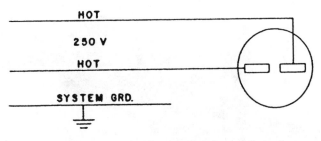

Fig. 4-18 *A 250-volt, 15-ampere tandem configuration.*

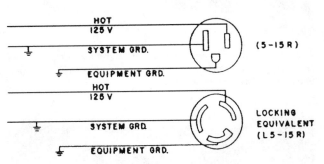

Fig. 4-19 *NEMA 125-volt, 15-ampere U-blade configuration.*

115-volt, single-phase applications that require a separate equipment ground. It is growing in popularity. It will probably replace the two-pole, two-wire configuration as the older type is phased out by replacements.

System ground   A system ground is the grounding of the neutral conductor, or ground leg, of the circuit to prevent lightning or other high voltages from exceeding the design limits of the circuit. It also limits the maximum potential to ground due to normal voltage.

Equipment ground   An equipment ground is the grounding of exposed conductive materials, such as conduit, switch boxes, or meter frames, which enclose conductors and equipment. This is to prevent the equipment from exceeding ground potential.

GROUNDING ADAPTERS

One of the problems with a grounding adapter is the lack of a grounded screw to which to attach the green pigtail at the plug. In Fig. 4-20, the ground wire has been added and attached to a grounded cold water pipe. In some cases, there is no ground wire in the cable that services the outlet. Therefore, there is no ground to the metal box. In some instances, the box is an insulated type made of nonmetallic materials. In other cases, the old knob-and-tube wiring may service the outlet or receptacle.

These are some of the reasons why the National Electrical Code (NEC) prohibits the use of such a pigtail adapter. They are, however, widely available and used. Refer to Fig. 4-21 to see why this could be a fatal

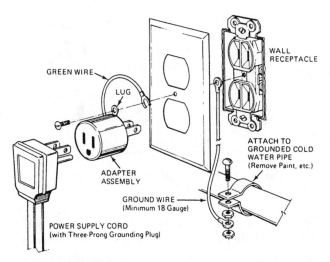

Fig. 4-20 *One method of connecting a ground wire to an ungrounded receptacle.*

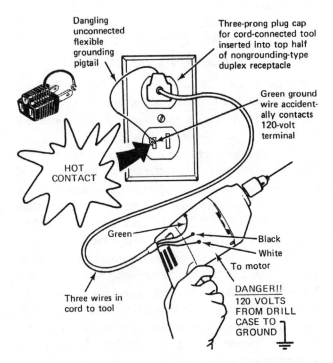

Fig. 4-21 *A pigtail adapter can cause problems if not properly attached to the screw.*

mistake. Once the pigtail adapter contacts the hot side of the outlet—it has not been attached to the faceplate screw since it would take time to do so—the handle of the drill is *hot* with 120 volts. If the person holding the drill touches a ground anywhere, the path for current flow is through the body.

One plug type is recognized by both the UL and the NEC (Fig. 4-22). The grounding adapter has a rigid tab with a spade end. This tab connects the adapter's grounding terminal to the metal screw contacting the grounded metal yoke that mounts the receptacle to

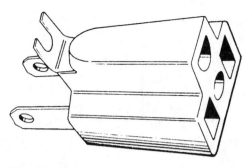

Fig. 4-22 *The UL- and NEC-recognized adapter.*

Fig. 4-23 *This is a typical residential ground using armored cable to positively ground the conductor from the service enclosure to the grounding electrode water pipe.*

Fig. 4-24 *This clamp bonds the jumpers around meters or other breaks. No additional hubs are needed.*

the grounded metal outlet box. If a nonmetallic outlet box is used, the adapter's's grounding terminal may be connected to an equipment grounding conductor in non-metallic (NM) cable. Different-width blades on the adapter permit insertion in only one way into the polarized blade opening on the receptacle.

Special Ground-Fault Protective Devices

Fuses and circuit breakers cannot distinguish between fault current and normal load current. A ground-fault interrupter, however, is insensitive to load current. It responds only to a ground-fault current.

An imbalance between a phase wire and neutral, or between phase wires of as little as 3 milliamperes, will cause a ground-fault interrupter to operate in less than 50 milliseconds. Portable and permanently installed devices are available. These will provide safe, portable-tool installation, independent of the ground return path.

The National Electrical Code recognizes the importance of a path to ground of continuous low impedance. No other problem is dealt with more thoroughly. The Code requires certain basic practices to be followed to make the equipment-grounding system safe. It recognizes the importance of Underwriters Laboratories listings of various connectors used in a system.

Code Requirements for Grounding Conductors

The National Electrical Code requires that a system grounding conductor be connected to any local metallic water piping system available on the premises, provided that the length of the buried water piping is a minimum of 10 feet. A system may be less than 10 feet long, or its electrical continuity may be broken by either disconnection or nonmetallic fittings. In these cases, it should be supplemented by the use of an additional electrode of a type specified by Section 250-83. Figures 4-23 and 4-24 illustrate the various types of connectors for grounding or bonding.

- *Fittings.* Whatever the type of grounding electrode, a system-grounding conductor should be connected through any one of a number of UL-listed clamps. Grounding wire may be bare, insulated, or armored conductor, or it may be in EMT (thin-wall conduit) or rigid conduit. Selection of the proper fitting depends upon the type of conductor and conduit and the type and diameter of the grounding electrode.

An approved clamp should be made of cast bronze or brass, or of plain or malleable cast iron. In attaching the ground clamp, care should be taken to make sure that any nonconductive protective coating, such as paint or enamel, is removed. Also, the surfaces must be clean, to ensure a good electrical connection.

- *Bonding jumpers.* Wherever electrical continuity of the grounding circuit may be interrupted, either by insulation or by disconnection, a bonding jumper is required by the Code. If a grounding attachment to metallic underground water piping is not made on the street side of a water meter, a bonding jumper must be used around the meter. This is needed to ensure an uninterrupted path to ground. Bonding jumpers must also be used to span other items likely to become disconnected, such as valves and service unions. In addition, they must be used to span insulating sections, such as plastic pipe, certain water softeners, or other insulating links. Bonding jumpers require the same size conductor as that of the grounding conductor which is run to water pipe, or other grounding electrode.

Service equipment Continuity at the service equipment is essential and must be ensured. The fittings shown in Figs. 4-25 and 4-26, when properly applied, will help ensure the continuity of ground. Connections of the fittings are made at the metallic enclosures by means of grounding wedges, locknuts, and bushings. They are installed so as to ensure continuity of ground. Thus, care must be exercised to select fittings that will maintain good contact. Bonding locknuts should be used if the raceway does not use the largest concentric knockout.

Fig. 4-25 *Grounding bushings must maintain good electrical contact with the enclosure. The lay-in lug provides unusual flexibility in positioning and fastening the grounding conductor.*

Fig. 4-26 *This grounding wedge works equally well on existing or new installations. This design reduces installing time while ensuring continuity of ground.*

Grounding bushings may have an insulated throat to protect the insulation of the current-carrying wire. However, this does not affect their function as a grounding connection. Good electrical contact is maintained with the metallic enclosure by the ground wire attached to the screw. Good electrical contact is also maintained with the grounding conduit. All threaded fittings must be made up wrench-tight, as specified by the Code.

The system-grounding conductor is attached to the white wire (the grounded circuit conductor) within the service equipment. It should run without joint or splice to the grounding electrode. Generally, attachment of the grounding conductor is made to a built-in terminal on the neutral bar. The neutral bar terminal may not accommodate the size of grounding conductor required. Then, a pressure terminal of the correct size must be bolted on and used for the system ground attachment.

- *Bonding.* To ensure electrical continuity of the grounding circuit, bonding is required at all conduit connections in service equipment. It is required also at points where any nonconductive coating exists that might impair such continuity.

- *Where bonding is required.* Bonding is required at connections between service raceways, service cable armor, and all service equipment enclosures containing service-entrance conductors (including meter fittings and boxes). It is required also at any conduit or armor that forms part of the grounding conductor to the service raceway.

When meter housings are mounted outdoors, separate from the service equipment, the neutral wire in the meter should run to ground.

A meter located inside a building should have the neutral wire bonded to the meter housing and attached to the grounded service conductor.

On rigid conduit, good ground connections will be ensured at the point where conduit meets the tapped hub or boss on the enclosure. If service-entrance cable is used, it is recommended that watertight connectors be installed, to protect the joint from moisture.

An iron-bodied, universal-type entrance cap is recommended where service conductors from supply enter the building. Such a fitting works equally well on rigid conduit or on thin-wall conduit with no threading. These caps are of weatherproof design and function to protect as well as to fasten conductors to the structure. These caps are available in aluminum.

An emphatic warning is repeated throughout the National Electrical Code: NEVER USE SOLDER FOR GROUNDING CONNECTIONS.

SAFETY DEVICES

As a power source, electricity can create conditions almost certain to result in bodily harm, property damage, or both. It is important for workers to understand the hazards involved when they are working around electrical power tools, maintaining electrical equipment, or installing equipment for electrical operation.

Overcurrent devices   These should be installed in every circuit. They should be of the same size and type to interrupt current flow when it exceeds the capacity of the conductor. Proper selection takes into account not only the capacity of the conductor, but also the rating of the power supply and potential short circuits.

Fuses   These must be chosen according to type and capacity to fill a specific need. It is recommended that a switch be placed in the circuit so that fuses can be deenergized before they are handled. Insulated fuse pullers should also be used. Blown fuses should be replaced by others of the same type. Fuses should never be inserted in a live circuit.

Circuit breakers   These have long been used in high-voltage circuits with large current capacities. Recently they have become more common in other types of circuits. Circuit breakers should be selected for a specific installation by qualified engineers. They should be checked regularly by experienced maintenance personnel.

Ground-fault circuit interrupter   This is a fast-operating circuit breaker that is sensitive to very low levels of current leakage to ground. The ground-fault interrupter (GFCI) is designed to limit electric shock to a current- and time-duration value below that which can produce serious injury. The unit operates *only* on line-to-ground fault currents, such as insulation leakage currents or currents likely to flow during accidental contact with a hot wire of a 120-volt circuit and ground. *It does not protect in the event of a line-to-line contact.* There are two main types of GFCIs:

1. The differential ground-fault interrupter, available with various modifications, has current-carrying conductors passing through the circular iron core of a doughnut-shaped differential transformer. As long as all the electricity passes through the transformer, the differential transformer is not affected, and will not trigger the sensing circuit. If a portion of the current flows to ground and through the fault-detector line, however, the flow of electricity through the sensing windings of the differential transformer causes the sensing circuit to open the circuit breaker. These devices can be arranged to interrupt a circuit for currents of as little as 5 mA flowing to ground.

2. Another design is the isolation-type ground-fault interrupter. This unit combines the safety of an isolation system with the response of an electronic sensing circuit. In this setup, an isolating transformer provides an inductive coupling between load and line. Both hot and neutral wires connect to the isolating transformer. There is no continuous wire between.

In the latter type of interrupter, a ground fault must pass through the electronic sensing circuit, which has sufficient resistance to limit current flow to as low as 2 mA. This is well below the level of human perception.

The ground-fault interrupters by Hubbell operate on the differential transformer principle. They monitor load currents flowing in the protected circuit, comparing current flowing to the load with current flowing from it. Leakage to ground (ground fault) appears as the difference between these currents. When this exceeds a preset level, the GFCI interrupts the circuit.

Figure 4-27 is a miniportable plug-in type of GFCI with a 120-volt, 15-ampere capability. Just plug it into the socket and then plug the device to be used on 120 volts and under 15 amperes into the GFCI. It guards against shock hazard. It is used with pumps, extension cords, and hand tools as well as many other applications in the home, office, or factory. Safety-minded electricians use this unit for on-the-job protection.

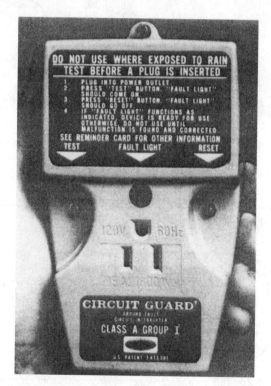

Fig. 4-27   *Mini portable, plug-in type ground-fault circuit interrupter (GFCI) for use on 120-volt, 15-ampere circuits.*

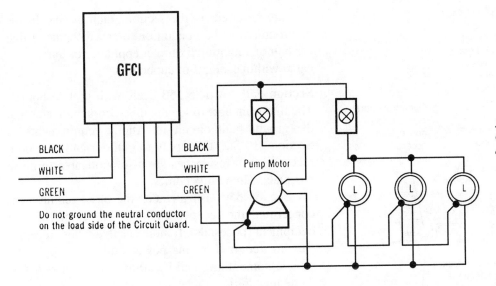

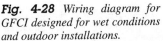

Fig. 4-28 *Wiring diagram for GFCI designed for wet conditions and outdoor installations.*

This device has a built-in test circuit that imposes an artificial ground fault on the load circuit to ensure that the ground-fault protection is functioning properly. This particular device trips at 5 mA.

Some GFCIs may be used in conjunction with underwater lights in older pools (built prior to 1965) which required less sensitive equipment, or one that has a 20-mA sensitivity. Some less sensitive types may also be used in industrial applications where moisture and chemicals preclude the use of the 5-mA type.

Figure 4-28 shows the wiring of a GFCI for use with a swimming pool. Pools require either a class A interrupter, which calls for a sensitivity of 5 mA to ground, or a class B model, which has a sensitivity of 20 mA to ground.

- The products included are intended primarily to protect normal* human beings from harmful effects of electric shock by sensing ground fault(s) and/or leakage current(s) on grounded and/or ungrounded systems rated 1000 volts AC or DC and below, and interrupting the electric circuit to the load when a fault current to ground and/or leakage current(s) exceeds some predetermined value that is less than that required to operate the overcurrent protective device of the supply circuit.

- The interrupting mechanism of this equipment may be separate from the sensing device or integral with the sensing device. Ground-fault circuit interrupters may be combined with other signaling or limiting products.

- Excluded are products primarily intended for ground-fault protection for equipment.

*The word *normal,* as used here, is intended to exclude persons who are electrically sensitive, either because of their unusual current conductivity or a physical defect.

A lightweight, compact GFCI is available. It has four outlets. One automatically trips when the supply voltage is interrupted or when the line voltage drops too low. It may be plugged into any 120-volt, 60-hertz line that has the proper overcurrent protection. The construction industry uses it where frequent damage to electrical equipment develops ground faults. Use of the portable GFCI can help protect an employer by preventing liability cases or claims.

GFCI-Related Code Requirements

This is a paraphrase summary of the various sections of the 2002 National Electrical Code that relate to ground-fault circuit interrupters:

Section 210-8(a) This requires that all 15- and 20-ampere, single-phase, 120-volt receptacles installed outdoors and in bathrooms of *all* residential occupancies be protected with ground-fault circuit interrupters (Fig. 4-29).

Section 210-8(b) This requirement makes it mandatory for all 15- and 20-ampere, 120-volt, single-phase receptacles on construction sites which are not part of the permanent wiring of the building or structure to be protected by GFCIs for personnel. An exception to this basic requirement is that receptacles on portable generators of 5 kW or less need not be protected (Figs. 4-30 and 4-31).

Section 210-52 This basic section tells you how many and where you have to put receptacle outlets. Of specific interest is the fact that it requires a receptacle outlet in the bathroom *adjacent to the basin location.* Further, it requires that there be at least one receptacle outlet installed outdoors on all one-family dwellings. Finally, this section indicates that the receptacle outlets

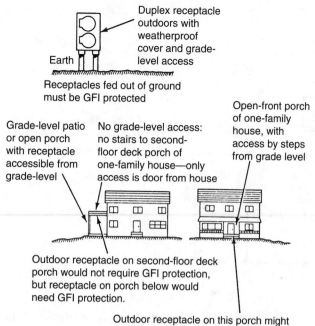

Duplex receptacle outdoors with weatherproof cover and grade-level access

Earth

Receptacles fed out of ground must be GFI protected

Grade-level patio or open porch with receptacle accessible from grade-level

No grade-level access: no stairs to second-floor deck porch of one-family house—only access is door from house

Open-front porch of one-family house, with access by steps from grade level

Outdoor receptacle on second-floor deck porch would not require GFI protection, but receptacle on porch below would need GFI protection.

Outdoor receptacle on this porch might be used to connect appliance used on ground, and GFI protection would seem to be required.

Fig. 4-29 *Receptacles fed out of ground must be GFCI-protected. Note locations of outdoor receptacles and those that require a GFCI device.*

that are required by the section, such as that in the bathroom, must be in addition to any receptacles that are part of a lighting fixture or appliance or that are located within cabinets or cupboards.

Section 550   Article 550 deals with mobile homes. The fine-print note following this section indicates that the ground-fault circuit interrupter requirements in Section 210-8(a) that apply to outdoor and bathroom receptacles in residences are also applicable to those receptacles in mobile homes.

Section 550 also includes the basic requirement concerning the receptacles in mobile home service equipment (pedestals). It indicates that where receptacles are required in this power outlet to supply the mobile home, they shall be 50-ampere receptacles. If there are additional receptacles in this power outlet of the 15- or 20-ampere, 120-volt, single-phase type, these additional receptacles shall be protected with GFCIs.

Section 680   Article 680 deals with swimming pools. This particular requirement says that no receptacle on the property shall be located within 10 feet of the inside walls of the pool, and that any receptacle located 10 to 15 feet from the inside walls shall be protected with a GFCI.

GFI circuit breaker protects all of the receptacles on this circuit.

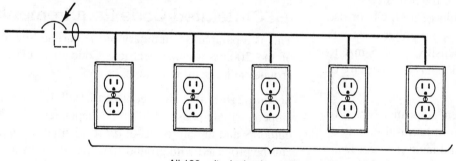

All 120-volt, single-phase, 15- and 20-ampere receptacle outlets connected to one or more branch circuits with GF I-CB protection

Fig. 4-30 *Ways that satisfy the basic rule on personnel shock protection on construction sites.*

Each receptacle assembly is a GFI type receptacle

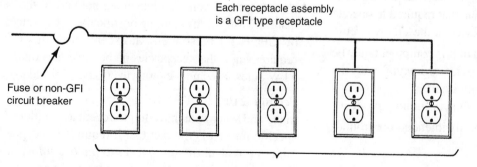

Fuse or non-GFI circuit breaker

All receptacles on construction site are GFI type

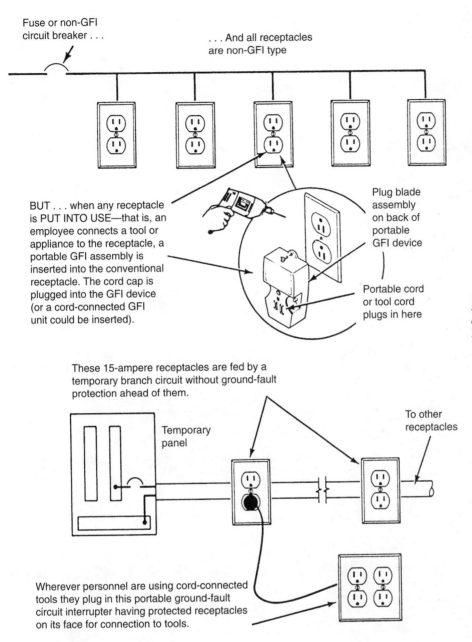

Fuse or non-GFI
circuit breaker . . .

. . . And all receptacles
are non-GFI type

BUT . . . when any receptacle
is PUT INTO USE—that is, an
employee connects a tool or
appliance to the receptacle, a
portable GFI assembly is
inserted into the conventional
receptacle. The cord cap is
plugged into the GFI device
(or a cord-connected GFI
unit could be inserted).

Plug blade
assembly
on back of
portable
GFI device

Portable cord
or tool cord
plugs in here

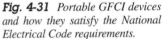

Fig. 4-31 *Portable GFCI devices and how they satisfy the National Electrical Code requirements.*

These 15-ampere receptacles are fed by a
temporary branch circuit without ground-fault
protection ahead of them.

Temporary
panel

To other
receptacles

Wherever personnel are using cord-connected
tools they plug in this portable ground-fault
circuit interrupter having protected receptacles
on its face for connection to tools.

Section 680-6(b)(1) This section pertains to lighting fixtures installed around swimming pools. It requires that any lighting fixture 5 feet horizontally from the inside walls of the pool be protected by a GFCI unless it is 12 feet above the maximum water level.

Section 680-6(b)(2) When a new swimming pool is installed next to an existing building with a lighting fixture on the building, it's possible for that existing lighting fixture to be within 5 feet of the inside edge of the pool. If this is the case, then the existing lighting fixture must be protected by a GFCI.

Section 680-6(b)(3) This section requires GFCI protection of any lighting fixture where the fixture or its

supporting means is located within 16 feet of any point on the water surface.

Section 680-20(a)(1) This requirement mandates the use of GFCIs to protect all electrical equipment, including the power supply cords used with storable (aboveground) swimming pools.

Section 680-50 This requirement applies to fountains. It requires that all electrical equipment such as lighting fixtures, submersible pumps, etc., be protected by GFCIs except where that equipment is operated below 15 volts.

5
CHAPTER

Wiring to the Electrical Code Requirements

ELECTRICITY HAS NOT BEEN AROUND TOO LONG in the relative span of time. Still, it is old enough to be considered an energy source no one would want to be without.

This was not always the case, however. Thomas Edison met with resistance from those who preferred gas lighting. To this day, we have some streets lighted by gas. But open flame, with its constant threat of uncontrolled fires, has always been a danger to people and their wooden buildings.

These are some examples of the destruction that fire has wrought: Property worth more than $15 million was destroyed in New York City when 13 acres (5.2 hectares) burned in 1835. Much of the town of Charleston, South Carolina, burned to the ground in 1861, with $10 million damage. In Portland, Maine, following the July 4 celebration in 1866, a $10 million blaze destroyed 1500 buildings. At today's prices this would be billions instead of millions.

Because these fires meant heavy losses to fire insurance companies, the companies began a cooperative effort to *prevent* fires. The New York Board of Fire Underwriters was formed in May 1867. The function of this organization was to arrange for inspection of premises. It was also to make recommendations for the removal of possible fire hazards. In 1882, the Committee of Surveys drew up a set of safeguards for *arc and incandescent lighting*. These safeguards were the forerunners of the present National Electric Code.

A letter from Thomas Edison to the New York Board of Fire Underwriters, in May 1881, shows how firmly he believed that his electric light company was "...free from any possible danger from fire." He believed his system of lighting was far superior to natural gas, that it provided protection from fire, and that it was safer in all regards.

As you know, electricity can cause shock and fire when improperly used or installed. This is why there are a number of codes written for the protection of those who generate and use electricity.

NATIONAL ELECTRICAL CODE (NEC)

The National Electrical Code was first developed in 1911. It is sponsored by the National Fire Protection Association (NFPA), an organization established in 1896.

The Code is reprinted every 3 years. There are a number of changes each time a new edition is issued. Additions or changes are agreed upon by a committee which, after studying recommendations from many sources, passes on their final form and wording before presenting them to the NFPA for publishing.

"This *Code* is purely advisory as far as the National Fire Protection Association and National Standards Institute are concerned, but it is offered for use in law and for regulatory purposes in the interest of life and protection of property."

The National Electrical Code serves as a basis on which local government authorities can write ordinances that deal with protection of the lives of people working with or using electricity or electrical devices. Local laws almost always refer to the National Electrical Code as the "standard minimum," sometimes making additions to it to meet local conditions.

For example, burying of metal in some types of soils can cause deterioration of the metal used for grounding, or for conduits, more quickly than for most other locations. This unique soil condition must therefore be pointed out in a *local* modification to the National Electric Code.

The Code is concerned with the *how* of electrical installation. It states the approved quantity and sizes of wire that can safely be used in a conduit as well as the number of wires in a box, and whether the wire should be aluminum or copper.

For example, No. 12 AWG aluminum wire, also copper-clad, can carry 15 amperes safely, if the types of wire are RUW, T, TW and if the temperature in which the wire is installed for operation does not exceed 60°C (140°F). This assumes that there are not more than three conductors in a raceway (conduit) or buried cable.

RUW means the insulation is made of moisture-resistant latex rubber and is suitable for dry and wet locations with a maximum operating temperature of 60°C (140°F). *T* means the insulation is thermoplastic, with the same maximum temperature, but for use in dry locations only. *TW* means the insulation is moisture-resistant thermoplastic, for the same maximum temperatures, and can be used in both dry and wet locations.

The National Electrical Code may be bought in paperback or as a hardcover book. It contains much of the information needed by an electrician on the job. It has definitions of electrical terms and chapters on wiring design and protection; wiring methods and materials; equipment for general use; special occupancies, equipment, and conditions; sections on communications; and tables, an index, and an appendix. A membership list of the National Electrical Code committee is included, stating titles held and addresses of individual members. The members come from all parts of the country. This gives greater input to Code revisions, keeping it current and, as nearly as possible, applicable to the entire country. The 1971 Code, for example, contains a very much revised article on swimming pools, fountains, spas or hot tubs, storable swimming pools, and wading pools (Article 680). The use of *ground-fault interrupters* is called for here to protect those around the pool as well as in the water. Grounding methods for electrical devices are stressed in the handbook.

No electrician should be without the latest edition of the Code book in his or her tool box. It has answers to most wiring problems. It may be obtained from:

National Fire Protection Association
Batterymarch Park
Quincy, Massachusetts 02269

UNDERWRITERS LABORATORIES

Underwriters Laboratories, Inc., was founded in 1894 by William Henry Merrill. He came to Chicago to test the installation of Thomas Edison's new incandescent electric light at the Columbian Exposition. He later started the UL for insurance companies to test products for electric and fire hazards. It continued as a testing laboratory for insurance underwriters until 1917. It then became an independent, self-supporting, safety-testing laboratory. The National Board of Fire Underwriters (now American Insurance Association) continued as sponsors of UL until 1968. At that time, sponsorship and membership were broadened to include representatives of consumer interests, government bodies or agencies, education, public safety bodies, public utilities, and the insurance industry, in addition to safety and standardization experts.

UL has expanded its testing services to more than 13,000 manufacturers throughout the world. Over one billion UL labels are used each year on products listed by Underwriters Laboratories.

UL is chartered as a not-for-profit organization without capital stock, under the laws of the state of Delaware. Its function is to establish, maintain, and operate laboratories for the examination and testing of devices, systems, and materials.

Its objectives, as stated by UL are as follows:

By scientific investigation, study, experiments and tests, to determine the relation of various materials, devices, products, equipment, constructions, methods and systems to hazards appurtenant thereto or to the use thereof, affecting life and property, and to ascertain, define and publish standards, classifications and specifications for materials, devices, equipment, construction, methods, and systems affecting such hazards, and other information tending to reduce and prevent loss of life and property from such hazards.

The corporate headquarters, together with one of the testing laboratories, is located at 333 Pfingsten Rd., Northbrook, Illinois 60062; at Melville, New York; at Santa Clara, California; and at Tampa, Florida.

Underwriters Laboratories, Inc., has a total staff of more than 2000 employees. More than 800 persons are engaged in engineering work at Melville. Of this number, approximately 425 are graduate engineers (Fig. 5-1). Supplementing the engineering staff are more than 500 factory inspectors in 70 countries worldwide.

Engineering functions of Underwriters Laboratories are divided among these six departments:

- Burglary protection and signaling
- Casualty and chemical hazards
- Electricity (Fig. 5-2)
- Fire protection (Fig. 5-3)
- Heating, air conditioning, and refrigeration
- Marine equipment

The electrical is the largest of the six engineering departments. Safety evaluation are made on hundreds of different types of appliances for use in homes, commercial buildings, schools, and factories. The scope of the work in this department includes electrical construction product in detail, particularly its construction and test performance. The procedure becomes the manufacturer's guide for future production. It is used by the UL inspectors for periodic review of the listed products. Procedures are kept up to date by the manu-

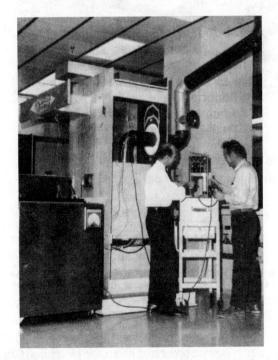

Fig. 5-1 *Testing furnaces and air conditioners at Underwriters Laboratories* (UL).

Fig. 5-2 *Checking electrical equipment at Underwriters Laboratories (UL).*

Fig. 5-3 *Testing fire extinguishers at UL. (UL)*

facturer, who informs the UL of any changes in the product or additions to the line (Fig. 5-4).

CANADIAN STANDARDS ASSOCIATION (CSA)
History of the CSA

The Canadian economy expanded after the turn of the twentieth century. It then became apparent that technical standardization was essential if Canada was to compete in the world markets. The original standards association was incorporated under the Dominion Companies Act in 1919, with the name Canadian Engineering

Fig. 5-4 *UL stickers and labels. (UL)*

Standards Association, a nonprofit, independent, voluntary organization.

By Supplementary Letters Patent, in 1944, this association became the Canadian Standards Association and extended its activities to a broader field of standardization.

Purpose of the Association

The basic objectives of the CSA are

- To develop voluntary national standards
- To provide certification services for national standards
- To represent Canada in international standards activities

Location and Operation of the Association

CSA headquarters and major testing facilities are located in Rexdale, a suburb of Toronto, in Ontario, Canada. These facilities house a staff of over 400, which includes professional engineers and skilled laboratory technicians in addition to management and secretarial personnel.

Regional offices and test facilities in Montreal, Winnipeg, and Vancouver are maintained for the convenience of clients across Canada. There is a branch office in Edmonton, Alberta. Executive offices are located in the national capital of Ottawa, Ontario.

CSA mark Products certified by CSA are eligible to bear the CSA certification mark. Misuse of the mark may result in suspension or cancellation of certification. CSA may resort to legal action to protect its registered trademark in the event of abuse (Fig. 5-5). In addition,

Fig. 5-5 *Canadian Standards Association symbol. (CSA)*

CSA information tags and other markings are made available for certified products and their containers, to supplement the CSA mark.

For your own safety, the CSA symbol on electrical equipment you purchase is an assurance that such equipment has passed rigid inspection.

WIRING ACCORDING TO THE NATIONAL ELECTRICAL CODE

There are a number of methods of wiring. Some use materials totally unacceptable in certain areas of the United States. However, local codes vary. This means the National Electrical Code must establish the standards for all the various types of wiring materials. Even the knob-and-tube method shown in Chapter 14, although obsolete, is still used in some areas of the United States. Therefore, the Code has established certain proper procedures in working with this type of wiring system. It can be added to, but it cannot be installed in any new construction.

BX cable, which is armored cable (AC), is specified in some locations. It is not permitted in other parts of the country. However, the Code has certain established rules that regulate its use.

Armored Cable

Type AC (armored cable) is familiarly used in all types of electrical systems for power and light branch circuits as well as feeders. Type AC is also used for signal and control circuit work. It is called for in some places because of its mechanical strength. The wires inside are protected by the metal covering (Fig. 5-6). The steel armor protects the wires. A bare grounding strip runs inside the armor for better equipment grounding.

Working with armored cable (AC)   Working with armored cable can be difficult. In sawing the armored cable, a hacksaw should be held as shown in Fig. 5-7. Once the armor has been cut with the hacksaw, a twist will remove the unwanted end and expose the wires needed for making a connection. See Fig. 5-8 for a different method of cutting armored cable. The Roto-Split cutter provides a clean cut and saves time as compared

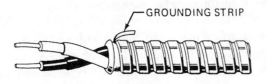

Fig. 5-6 Armored cable. Steel armor is wound around the wires for protection. Note the grounding wire.

GROUNDING STRIP

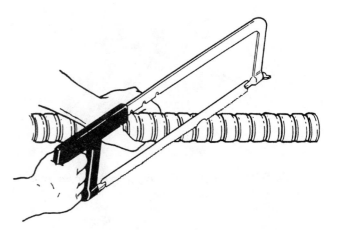

Fig. 5-7 *Cutting armored cable with a hacksaw.*

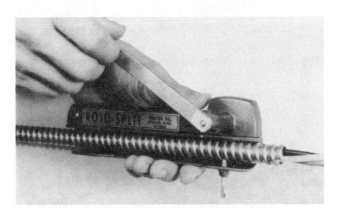

Fig. 5-8 *The rotor split cutter for armored cable. (Seatek)*

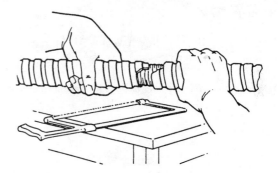

Fig. 5-9 *Twisting the cut end of armor to be removed.*

with the hacksaw. Figure 5-9 shows what happens once the cut cable is twisted.

After twisting you must unwrap the paper over the wires as far as you can. Do not tear it off (Fig. 5-10). Remove the paper from inside the armor as far as you can. Then, with a sharp yank, tear it off *inside* the armor (Fig. 5-11). Once the paper has been removed, you must insert a fiber bushing around the armor and wires (Figs. 5-12 and 5-13). This protects the wires against damage. Their insulation could be punctured by the cut ends of the metal cable housing during installation.

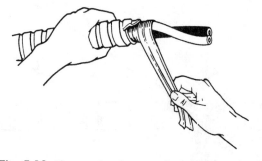

Fig. 5-10 *Unwrapping the paper from the exposed end.*

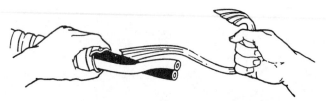

Fig. 5-11 *Snapping the paper to break it off inside the armored cable.*

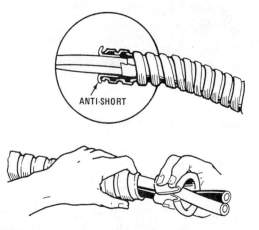

Fig. 5-12 *Inserting the fiber bushing for antishorting purposes.*

Fig. 5-13 *The fiber bushing used to protect against damage to the wires in armored cable.*

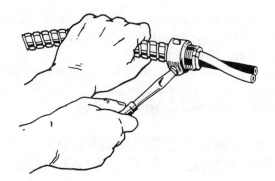

Fig. 5-14 *Placing a connector on armored cable.*

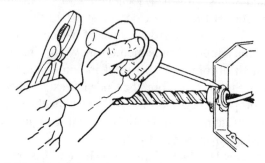

Fig. 5-15 *Driving the locknut in place within the box to hold the connector of the armored cable securely.*

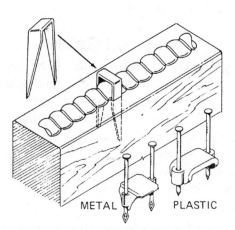

Fig. 5-16 *Staples are used to support armored cable on wooden surfaces.*

Installing a connector on the armored cable can be done as shown in Fig. 5-14. The connector can then be placed inside a box as shown in Fig. 5-15. Drive the locknut of the connector down tightly. Make sure the teeth of the locknut bite into the metal of the box.

Armored cable must be supported every 4½ feet by a method such as that shown in Fig. 5-16. In addition to this support, it should be supported within 12 inches of each box. However you can leave a free length not over 24 inches long at terminals where flexibility is needed. This may be the case with a motor that must be adjusted for belt tension. Keep in mind that the staples used must be rustproof. The NEC requires it.

Figure 5-17 shows how the BX cable is attached to the inside of a metal box for use in hooking up a combination duplex receptacle. This one will accommodate a 250-volt tandem plug and a 125-volt standard three-prong plug. Note the BX has four wires.

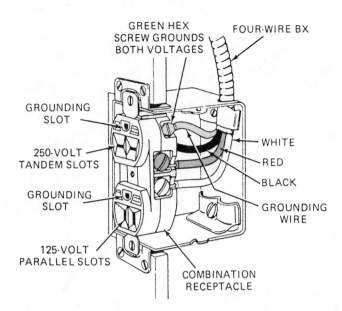

Fig. 5-17 *Placement of armored cable within a box and connecting it to a duplex receptacle.* (Harvey Hubbell)

Fig. 5-18 *Rigid metal conduit.*

Rigid Metal Conduit

This method of wiring involves pipes that enclose the wires. The conduit is a pipe. It is aluminum or steel. This is not the same as water pipe. The inside of the pipe must be smooth so the wires can be pulled through without ruining the insulation. The steel pipe bends easily. The type of steel used allows for manual bending. The conduit may be black. If so, it is used indoors only in dry locations. The conduit also may be galvanized.

Some aluminum conduit and some steel conduit have additional coatings of plastic to protect them. Aluminum conduit must have an additional plastic coating if it is to be used in contact with the earth or buried in concrete.

Conduit has a UL label (Fig. 5-18). All sizes are identical in dimensions with the corresponding size of water pipe, but the nominal size indicates the internal diameter with water pipe. However, the actual diameter is larger than that of the water pipe. Conduit, for instance, of ½-inch diameter will actually have a diameter closer to ⅝ inch. The actual outside or external diameter will be about ⅞ inch.

Use a special conduit cutter for reducing the conduit from its usual 10-foot length. A hacksaw will work. Do not use the pipe cutter. A pipe cutter will leave a sharp edge inside the conduit. This edge will scrape the insulation and cause shorts in the electrical wiring. A reamer should be used to make sure the cut surfaces are smooth.

Figure 5-19 shows how a piece of rigid conduit is used to support a box that has a duplex receptacle. Note the grounding arrangement in addition to the pipe itself.

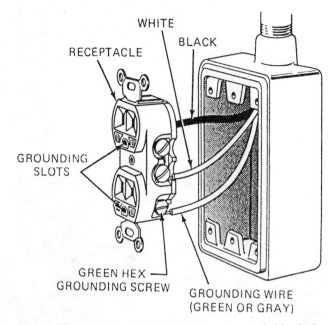

Fig. 5-19 *Placement of rigid conduit into a box to hold a duplex receptacle.* (Harvey Hubbell)

Figure 5-20 shows the use of rigid conduit for a residential installation of the service entrance. Note the straps used to hold the rigid conduit securely.

Figure 5-21 shows how a strap is used to support a piece of conduit near the coupling and near the connector.

The number of conductors in a piece of conduit is regulated by the Code. The number of conductors permitted in a particular size of conduit or tubing is covered in Chapter 9 of the Code (Tables 5-1 and 5-2). These tables cover conductors of the same size used for either new work or rewiring. Tables 5-3 and 5-4 cover combinations of conductors of different sizes when used for new or rewiring. In most cases the Code does not permit more than 40% of the conduit area to be filled with wires.

A conduit run should be installed so that no injury comes to the wires being pulled through it. The basic rule for support of rigid conduit is within every 3 feet of a connection to an enclosure or fitting and then every 10 feet. If threaded couplings are used, then wider spacing for supports can be used.

Nonmetallic Conduit

Plastic conduit is called nonmetallic. This plastic conduit is listed in sizes from ½ inch to 6 inches. It is

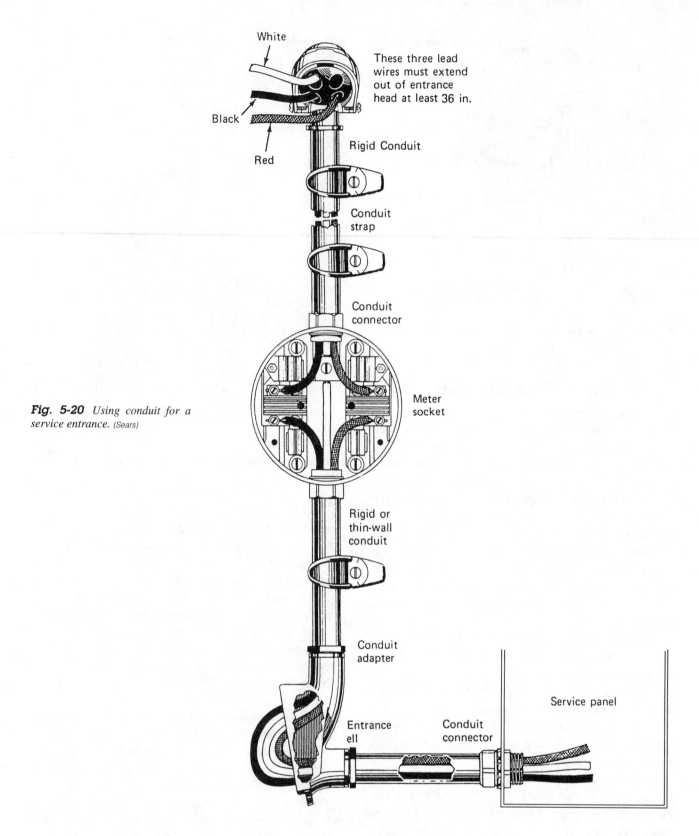

White

These three lead wires must extend out of entrance head at least 36 in.

Black

Red

Rigid Conduit

Conduit strap

Conduit connector

Fig. 5-20 *Using conduit for a service entrance.* (Sears)

Meter socket

Rigid or thin-wall conduit

Conduit adapter

Service panel

Entrance ell

Conduit connector

used primarily where there may be a corrosive atmosphere that would damage metal conduit. It can be used aboveground for circuits operating with voltages up to 600 volts. It can be used as a general-purpose raceway. Only polyvinyl chloride (PVC) is acceptable for use aboveground in buildings.

It should be kept in mind that nonmetallic conduit is not permitted in ducts, plenums, and other air handling spaces.

Support rules for nonmetallic conduit are simple and direct. A clamp or other support must be placed within 3 feet of any box, cabinet, or other conduit termination.

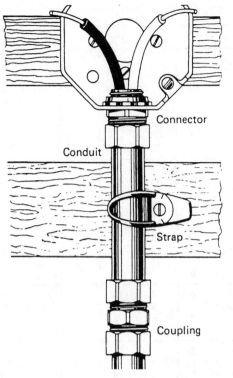

Fig. 5-21 *Note the support strap used near the connector and coupling.*

Spacing between supports varies with the temperature rating and the size of the conduit. See Section 347 of the Code for exact support spacing. It may vary from 2 to 8 feet.

All cuts should be trimmed inside and out, and the surfaces made smooth. Couplings and enclosures must be of the type approved for the conduit. In plastic conduit the cement used to make connections should also be of the type approved by the Code or local ordinance.

Electrical Metallic Tubing (EMT)

This type of conduit is thinner in outside wall thickness than rigid conduit. The internal diameter in the smaller sizes is the same as that in rigid conduit. In the larger sizes, the internal diameter is a little larger. The walls are thin and should not be threaded. All fittings are threadless. They hold to the material through pressure. Figure 5-22 shows a piece of EMT. Figure 5-23 shows a connector and a coupler for use with thin-wall conduit.

To cut thin-wall conduit, use a hacksaw with 32 teeth to the inch. A metal-removing power saw can also be used. The burrs and sharp edges should be removed

Table 5-1 *Maximum Number of Conductors in Trade Sizes of Conduit or Tubing**

Type Letters	Conductor Size AWG, MCM†	½	¾	1	1¼	1½	2	2½	3	3½	4	4½	5	6
TW, T, RUH, RUW, XHHW (14 through 8)	14	9	15	25	44	60	99	142						
	12	7	12	19	35	47	78	111	171					
	10	5	9	15	26	36	60	85	131	176				
	8	2	4	7	12	17	28	40	62	84	108			
RWH and RHH (without outer covering), THW	14	6	10	16	29	40	65	93	143	192				
	12	4	8	13	24	32	53	76	117	157				
	10	4	6	11	19	26	43	61	95	127	163			
	8	1	3	5	10	13	22	32	49	66	85	106	133	
TW, T, THW, RUH (6 through 2), RUW (6 through 2)	6	1	2	4	7	10	16	23	36	48	62	78	97	141
	4	1	1	3	5	7	12	17	27	36	47	58	73	106
	3	1	1	2	4	6	10	15	23	31	40	50	63	91
	2	1	1	2	4	5	9	13	20	27	34	43	54	78
	1		1	1	3	4	6	9	14	19	25	31	39	57
FEPB (6 through 2), RHW and RHH (without outer covering)	0		1	1	2	3	5	8	12	16	21	27	33	49
	00		1	1	1	3	5	7	10	14	18	23	29	41
	000		1	1	1	2	4	6	9	12	15	19	24	35
	0000			1	1	1	3	5	7	10	13	16	20	29
	250			1	1	1	2	4	6	8	10	13	16	23
	300			1	1	1	2	3	5	7	9	11	14	20
	350				1	1	1	3	4	6	8	10	12	18
	400				1	1	1	2	4	5	7	9	11	16
	500				1	1	1	1	3	4	6	7	9	14
	600					1	1	1	3	4	5	6	7	11
	700					1	1	1	2	3	4	5	7	10
	750					1	1	1	2	3	4	5	6	9

*Reprinted by permission from NFPA, *National Electrical Code,* © 2002 National Fire Protection Association.

†MCM = 1,000 circular mils.

Table 5-2 *Maximum Number of Conductors in Trade Sizes of Conduit or Tubing**

Type Letters	Conductor Size AWG, MCM†	1/2	3/4	1	1¼	1½	2	2½	3	3½	4	4½	5	6
THWN	14	13	24	39	69	94	154							
	12	10	18	29	51	70	114	164						
	10	6	11	18	32	44	73	104	160					
	8	3	5	9	16	22	36	51	79	106	136			
THHN FEP (14 through 2) FEPB (14 through 8)	6	1	4	6	11	15	26	37	57	76	98	125	154	
	4	1	2	4	7	9	16	22	35	47	60	75	94	137
	3	1	1	3	6	8	13	19	29	39	51	64	80	116
	2	1	1	3	5	7	11	16	25	33	43	54	67	97
	1		1	1	3	5	8	12	18	25	32	40	50	72
XHHW (4 though 500 MCM)	0		1	1	3	4	7	10	15	21	27	33	42	61
	00		1	1	2	3	6	8	13	17	22	28	35	51
	000		1	1	1	3	5	7	11	14	18	23	29	42
	0000		1	1	1	2	4	6	9	12	15	19	24	35
	250			1	1	1	3	4	7	10	12	16	20	28
	300			1	1	1	3	4	6	8	11	13	17	24
	350			1	1	1	2	3	5	7	9	12	15	21
	400				1	1	1	3	5	6	8	10	13	19
	500				1	1	1	2	4	5	7	9	11	16
	600				1	1	1	1	3	4	5	7	9	13
	700					1	1	1	3	4	5	6	8	11
	750					1	1	1	2	3	4	6	7	11
XHHW	6	1	3	5	9	13	21	30	47	63	81	102	128	185
	600				1	1	1	1	3	4	5	7	9	13
	700					1	1	1	3	4	5	6	8	11
	750					1	1	1	2	3	4	6	7	10
RHW	14	3	6	10	18	25	41	58	90	121	155			
	12	3	5	9	15	21	35	50	77	103	132			
	10	2	4	7	13	18	29	41	64	86	110	138		
	8	1	2	4	7	9	16	22	35	47	60	75	94	137
RHH (with outer covering)	6	1	1	2	5	6	11	15	24	32	41	51	64	93
	4	1	1	1	3	5	8	12	18	24	31	39	50	72
	3	1	1	1	3	4	7	10	16	22	28	35	44	63
	2		1	1	3	4	6	9	14	19	24	31	38	56
	1		1	1	1	3	5	7	11	14	18	23	29	42
	0		1	1	1	2	4	6	9	12	16	20	25	37
	00			1	1	1	3	5	8	11	14	18	22	32
	000			1	1	1	3	4	7	9	12	15	19	28
	0000			1	1	1	2	4	6	8	10	13	16	24
RHW	250				1	1	1	3	5	6	8	11	13	19
	300				1	1	1	3	4	5	7	9	11	17
	350				1	1	1	2	4	5	6	8	10	15
	400					1	1	1	3	4	6	7	9	14
RHH (with outer covering)	500				1	1	1	1	3	4	5	6	8	11
	600					1	1	1	2	3	4	5	6	9
	700						1	1	1	3	3	4	6	8
	750						1	1	1	3	3	4	5	8

*Reprinted by permission from NFPA, *National Electrical Code,* © 2002 National Fire Protection Association.
†MCM = 1,000 circular mils.

with a reamer. A complete set of fittings and methods of using EMT can be seen in Chap. 11 in this book.

Flexible Metal Conduit

EMT is restricted in its use to where it will not be exposed to physical abuse or corrosive agents. Flexible metal conduit is also designed for use where it will not be subject to physical damage (Fig. 5-24).

A typical use of flexible metallic tubing is a 4- to 6-foot length for a fixture whip in a ceiling (Fig. 5-25). The conduit may contain two No. 18 Type AF wires for a 6-ampere fixture load. Section 240-4 of the Code

Table 5-3 Dimensions and Percent Area of Conduit and of Tubing*

Tables 5-3 and 5-4 give the nominal size of conductors and conduit or tubing for use in computing the size of conduit or tubing for various combinations of conductors. The dimensions represent average conditions only. Variations will be found in dimensions of conductors and conduits of different manufacture.

		Area (Square Inches)								
			Not Lead-Covered			Lead-Covered				
Trade Size	Internal Diameter (Inches)	Total 100%	2 Cond. 31%	Over 2 Cond. 40%	1 Cond. 53%	1 Cond. 55%	2 Cond. 30%	3 Cond. 40%	4 Cond. 38%	Over 4 Cond. 35%
1/2	0.622	0.30	0.09	0.12	0.16	0.17	0.09	0.12	0.11	0.11
3/4	0.824	0.53	0.16	0.21	0.28	0.29	0.16	0.21	0.20	0.19
1	1.049	0.86	0.27	0.34	0.46	0.47	0.26	0.34	0.33	0.30
1 1/4	1.380	1.50	0.47	0.60	0.80	0.83	0.45	0.60	0.57	0.53
1 1/2	1.610	2.04	0.63	0.82	1.08	1.12	0.61	0.82	0.78	0.71
2	2.067	3.36	1.04	1.34	1.78	1.85	1.01	1.34	1.28	1.18
2 1/2	2.469	4.79	1.48	1.92	2.54	2.63	1.44	1.92	1.82	1.68
3	3.068	7.38	2.29	2.95	3.91	4.06	2.21	2.95	2.80	2.58
3 1/2	3.548	9.90	3.07	3.96	5.25	5.44	2.97	3.96	3.76	3.47
4	4.026	12.72	3.94	5.09	6.74	7.00	3.82	5.09	4.83	4.45
4 1/2	4.506	15.94	4.94	6.38	8.45	8.77	4.78	6.38	6.06	5.56
5	5.047	20.00	6.20	8.00	10.60	11.00	6.00	8.00	7.60	7.00
6	6.065	28.89	8.96	11.56	15.31	15.89	8.67	11.56	10.98	10.11

*Reprinted by permission from NFPA, *National Electrical Code,* © 2002 National Fire Protection Association.

Table 5-4 Properties of Conductors*

Tables 5-3 and 5-4 give the nominal size of conductors and conduit or tubing for use in computing the size of conduit or tubing for various combinations of conductors. The dimensions represent average conditions only. Variations will be found in dimensions of conductors and conduits of different manufacture.

		Concentric Lay Stranded Conductors		Bare Conductors		DC Resistance (Ohms/Mft at 25°C (77°F))		
						Copper		
Size AWG MCM**	Area (Circular Mils)	No. Wires	Diam. Each Wire (Inches)	Diam. (Inches)	Area† (Square Inches)	Bare Cond.	Tin'd Cond.	Aluminum
18	1,620	Solid	0.0403	0.0403	0.0013	6.51	6.79	10.7
16	2,580	Solid	0.0508	0.0508	0.0020	4.10	4.26	6.72
14	4,110	Solid	0.0641	0.0641	0.0032	2.57	2.68	4.22
12	6,530	Solid	0.0808	0.0808	0.0051	1.62	1.68	2.66
10	10,380	Solid	0.1019	0.1019	0.0081	1.018	1.06	1.67
8	16,510	Solid	0.1285	0.1285	0.0130	0.6404	0.659	1.05
6	26,240	7	0.0612	0.184	0.027	0.410	0.427	0.674
4	41,740	7	0.0772	0.232	0.042	0.259	0.269	0.424
3	52,620	7	0.0867	0.260	0.053	0.205	0.213	0.336
2	66,360	7	0.0974	0.292	0.067	0.162	0.169	0.266
1	83,690	19	0.0664	0.332	0.087	0.129	0.134	0.211
0	105,600	19	0.0745	0.372	0.109	0.102	0.106	0.168
00	133,100	19	0.0837	0.418	0.137	0.0811	0.0843	0.133
000	167,800	19	0.0940	0.470	0.173	0.0642	0.0668	0.150
0000	211,600	19	0.1055	0.528	0.219	0.0509	0.0525	0.0836
250	250,000	37	0.0822	0.575	0.260	0.0431	0.0449	0.0708
300	300,000	37	0.0900	0.630	0.312	0.0360	0.0374	0.0590
350	350,000	37	0.0973	0.681	0.364	0.0308	0.0320	0.0505
400	400,000	37	0.1040	0.728	0.416	0.0270	0.0278	0.0442
500	500,000	37	0.1162	0.813	0.519	0.0216	0.0222	0.0354
600	600,000	61	0.0992	0.893	0.626	0.0180	0.0187	0.0295
700	700,000	61	0.1071	0.964	0.730	0.0154	0.0159	0.0253
750	750,000	61	0.1109	0.998	0.782	0.0144	0.0148	0.0236
800	800,000	61	0.1145	1.030	0.833	0.0135	0.0139	0.0221
900	900,000	61	0.1215	1.090	0.933	0.0120	0.0123	0.0197
1000	1,000,000	61	0.1280	1.150	1.039	0.0108	0.0111	0.0177
1250	1,250,000	91	0.1172	1.289	1.305	0.00863	0.00888	0.0142
1500	1,500,000	91	0.1284	1.410	1.561	0.00719	0.00740	0.0118
1750	1,750,000	127	0.1174	1.526	1.829	0.00616	0.00634	0.0101
2000	2,000,000	127	0.1255	1.630	2.087	0.00539	0.00555	0.00885

*Reprinted by permission from NFPA, *National Electrical Code,* © 2002 National Fire Protection Association.

**MCM = 1,000 circular mils.

†Area given is that of a circle having a diameter equal to the overall diameter of a stranded conductor.

Fig. 5-22 *Electrical metallic tubing.*

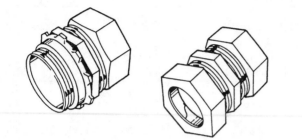

Fig. 5-23 *A connector and a coupling used with thin-wall conduit (EMT).*

Fig. 5-24 *Flexible metal conduit. Wires are pulled through this conduit as with any other type.*

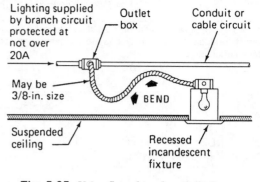

Fig. 5-25 *Using flex tubing for a light fixture.*

allows No. 16 and No. 18 fixture wire to be protected at 20 amperes. The flexible tubing is also the equipment grounding conductor. It should be limited to dry locations if it does not have a nonmetallic coating. It cannot be used for direct earth burial or for embedding in poured concrete or aggregate. It is also limited to lengths of not more than 6 feet. This means it is ruled out from being used as a general-purpose raceway. It has not been judged as being adequate for grounding purposes for lengths over 6 feet. The 3/8-inch flexible tubing usually can handle two No. 14 Type AF wires and no grounding conductor. It is primarily used for attaching light fixtures. Straps or supports should be placed within 12 inches of each end and with 4½ feet in between. Unclamped lengths of flex are permitted by an exception to the Code.

For concealed work, the run of flex from one outlet to another or to a fitting must not have more than the equivalent of four quarter-bends. This includes those bends located at the outlet or fitting. Angle connectors cannot be used for concealed flex installations. Straight connectors, however, can be used.

6
CHAPTER

Planning

IN NEW CONSTRUCTION, THE RESIDENTIAL electrician has to follow the plans given to him or her. Any changes must be cleared with the subcontractor, the contractor, and the owner. In remodeling, the electrician is working, in most cases, directly with the owner. Decisions can be made at the time a problem arises. However, there are some very basic things to be considered in making a plan a reality. There are code restrictions. Codes are local in nature and require various types of hookups, wire sizes, and boxes. The method of installation is very important for the safety of the resident.

The National Electric Code (NEC) is used as a basic standard when there is no other code to be followed. In this chapter we deal with the requirements for a good living environment. At the same time, we point out some of the Code restrictions that should be used as a basis for wiring the house.

Planning is important so that the proper amount of materials and equipment is handy when needed. Once the job is finished, the work should pass inspection by a person other than the one who installed it. In most communities the local electrical inspector is called to check the wiring before it is enclosed behind drywall and after it is finished and operating as a system.

The correct wire size, from the transformer on the pole to the outlet on the wall, helps prevent such occurrences as blinking lights when the refrigerator turns on. Wire size is determined by the amount of current required by power-consuming devices. These devices present a load to the power source. The load is usually shown in watts on appliance nameplates. Power, which is measured in watts, is determined by multiplying the voltages by the current. If an overload exists, or if the wires are too small, a voltage drop caused by the small wires reduces power available inside the house.

The larger the physical size of the wire, the more current it can handle without line losses caused by an increase in the load. Wire size is stated in numbers. The larger the number, the smaller the wire.

FIGURING CIRCUIT LOAD

The load presented by a circuit may be figured by simply adding the wattage ratings of the devices to be attached to the line or plugged into the circuit. When you are planning a new house or rewiring an old one, it is best to figure the amount of wattage needed and then design the circuits to accommodate the load.

There are a number of reasons why you need adequate wiring in a house—new or old. If fuses blow, or a circuit breaker trips, probable overloads exist. If the lights dim, blink, or flicker whenever an additional device is turned on, then the circuits are overloaded. A television picture will shrink if it is on a line with an overload. If it is a color television, the colors may change when something else turns on in the house. If an electric iron or electric range takes longer than usual to heat up, the chances are the circuits are overloaded. Air conditioners place quite a load on available circuitry. If the air conditioner is more than 5000 Btu, it should be on a separate circuit.

Rewiring may be necessary to eliminate some of these problems. However, at the time a new house is being planned, it is best to make sure future electrical needs are taken into consideration. During the past 20 years, the use of electricity in the home has more than tripled. Probably, the future will hold more demands for electricity. Adequate planning is a must if blinking lights, shrinking TV pictures, and blown fuses are to be prevented.

General Requirements for Residential Wiring

A fundamental prerequisite for any adequate wiring installation is conformity to the safety regulations applicable to the dwelling. The approved U.S. standard for electrical safety is the National Electrical Code. Usually, local safety regulations in municipal ordinances or state laws are based on the National Electrical Code.

Inspection service, to determine conformance with safety regulations, is available in most communities. Where available, a certificate of inspection should be obtained. In the absence of inspection service, a certificate should be obtained from the electrical contractor who installed the wiring. The certificate should say the wiring conforms with the applicable safety regulations.

Locating lighting outlets Proper illumination is an essential element of modern living. The amount and type of illumination required should be adapted to the various tasks in the home. There must also be planned lighting for recreation and entertainment. In many instances, best results are achieved by a blend of lighting from both fixed and portable light sources. Good home lighting, therefore, requires thoughtful planning and careful selection of fixtures, portable lamps, and other lighting equipment.

Where lighting outlets are mentioned in plans, their types and locations should conform with the lighting fixtures or equipment to be used. Unless a specified location is stated, lighting outlets may be located anywhere within the areas under construction.

Locating convenience outlets Convenience outlets should be located near the ends of wall space, rather than near the center. This reduces the likelihood of their being concealed behind large pieces of furniture. Unless otherwise specified, outlets should be located approximately 12 inches above the floor line.

Locating wall switches Wall switches should normally be located at the latch side of the doors, or at the traffic side of arches, and within the room or area to which the control is applicable. Some exceptions to this practice are (1) control switches for exterior lights from indoors, (2) control of stairway lights from adjoining areas where stairs are closed off by doors at head or foot, and (3) the control of lights from the access space adjoining infrequently used areas, such as storage rooms. Wall switches are normally mounted approximately 48 inches above the floor.

Controlling from more than one switch All spaces for which wall switch controls are required, and which have more than one entrance, should be equipped with a multiple-switch control at each principal entrance. If this requirement would result in the placing of switches controlling the same light within 10 feet of each other, one of the switch locations may be eliminated.

Some rooms are lighted from more than one source, as when both general and supplementary illumination are provided from fixed sources. In this case, multiple switching is required for only one set of controls, usually to the general illumination circuit.

Principal entrances are those commonly used for entry to, or exit from, a room where one is going from a lighted to an unlighted condition, or the reverse. For instance, a door from a living room to a porch is a principal entrance to the porch. However, it would not necessarily be considered a principal entrance to the living room, unless the front entrance to the house were through a porch.

Number of Local Lighting or Convenience Outlets Necessary

If the standards require a certain number of outlets based on a linear or square-foot measure, the number of outlets shall be determined by dividing the total linear or square footage by the required distance or area. The numbers so determined should be increased by 1 if a major fraction remains.

Example

Required: One outlet for each 150 square feet

Total square feet of area: 390

390 divided by 150 equals 2.6

Outlets required: 3

Dual-purpose rooms A room may be intended to serve more than one function, such as a combination living-dining room or a kitchen-laundry. In this case, convenience and special-purpose outlet provisions of these standards are separately applicable to those respective areas. Lighting outlet provisions may be combined in any manner that will ensure general, overall illumination, as well as local illumination of work surfaces. In locating wall switches, a dual-purpose area is considered a single room.

Dual functions of outlets In any instance where an outlet is located to satisfy two different provisions of the standard of adequacy, only one outlet need be installed at that location. In such instances, particular attention should be paid to required wall switch controls, as additional switching may be necessary.

For example, a lighting outlet in an upstairs hall may be located at the head of the stairway. This satisfies both a hall-lighting outlet and a stairway-lighting outlet provision of Code standards with a single outlet. Stairway-lighting provisions will necessitate multiple-switch control of this outlet at both the head and the foot of the stairway. In addition, if the hallway is long, the multiple-switch control rule, previously stated, may require a third point of control in the hall.

INTERPRETING REQUIREMENTS

The following standards are necessarily general in nature. They apply to most situations encountered in normal house construction and are to be considered as *minimum* standards of adequacy for such construction. Situations will arise, because of unusual design or unusual construction methods, when it will be impossible to satisfy a particular provision of the standards. The following rules may serve as a guide for meeting such situations.

- Wiring installation should be fitted to the structure. If compliance with a particular provision would require alteration of doors, windows, or structural members, alternate wiring provisions should be made. For example, some types of building construction may make it desirable to alter the recommended outlet or switch heights presented here, or to resort to surface-type wiring construction.

- Each provision of the adequacy standard is intended to provide for one or more specific uses of electricity. If such use is appropriate to the particular home under study and cannot be provided in accordance with the standard, an alternate provision should be made. If a construction peculiarity eliminates the need for a particular electrical facility, no alternate provision is needed.

- If certain functions are omitted from the plans, such as a workshop or a laundry, electrical wiring serving these functions naturally would be omitted.

- If certain facilities are indicated as future additions to the plans, for example, a basement recreation room, the initial wiring should be so arranged that none of it need be replaced, or moved, when the ultimate plan is realized. Final wiring for the future addition may be left to a later date, if desired.

- Standards given here may be applied to multifamily dwelling units. At present there are no nationally recognized standards for common-use spaces in such buildings. This makes good judgment on the part of the planner of electrical wiring a necessity. Particular attention should be paid to limiting voltage drop in feeders to individual apartments to ensure proper operation of appliance and space-heating equipment.

TYPICAL OUTLET REQUIREMENTS FOR A HOME

Figures 6-1 through 6-5 show a method of drawing in outlets and switches on the floor plan of a house. The symbols used are identified in Table 6-1. The use of these symbols makes it easier to sketch in the requirements before the building is started. The homeowner can then check these requirements. In the case of remodeling, the symbols can be used to show where new wiring and new electrical devices will be installed. This preplanning makes it much easier for the electrician to figure out the amount of wire and the devices needed for a job. Such preplanning saves time and energy.

Other considerations in the wiring plan of a house include doorbells, chimes, communications centers (intercoms), household fire-warning systems, and televisions with special lead-ins and rotor wire. These items are discussed in other chapters of this book.

Planning Requirements for Individual Areas

Exterior entrance wiring   The following planning requirements should be observed.

Lighting provisions   It is recommended that lighting outlets, controlled from wall switches, be installed at all entrances (Fig. 6-6). The principal lighting requirements for entrances are the illumination of steps leading to the entrance and of faces of people at the door.

Convenience outlets   Outlets in addition to those at the door are often desirable for post lights to illuminate terraced or broken flights of steps, or long approach walks. These outlets should be wall-switch-controlled from inside the house, near the entrance.

It is recommended that an exterior outlet near the front entrance be controlled by a wall switch inside the entrance. This provides for convenient operation of outdoor lighting, lawn mowers, and hedge trimmers.

Code requirements   The NEC requires a ground-fault interrupter (GFCI) if the outlet near the front entrance is outside. The rule requires GFCI protection if the receptacle is where there is direct grade-level

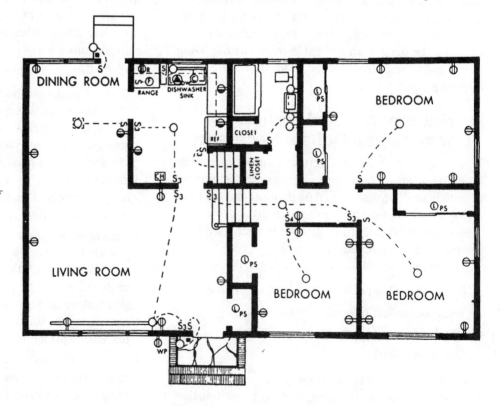

Fig. 6-1 *Wiring requirements of a typical home.*

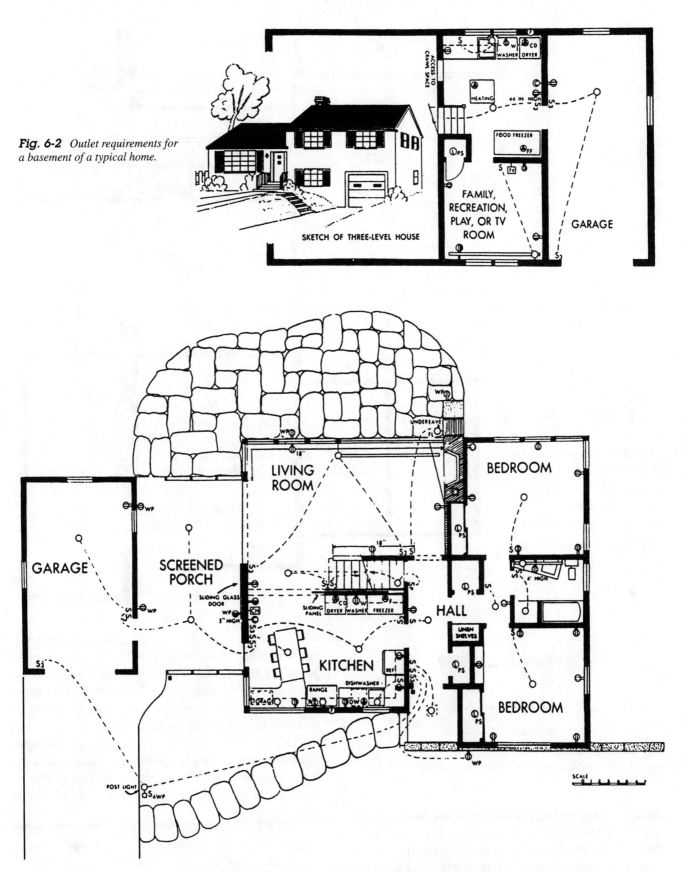

Fig. 6-2 *Outlet requirements for a basement of a typical home.*

Fig. 6-3 *Electrical wiring diagram for a floor of a home.*

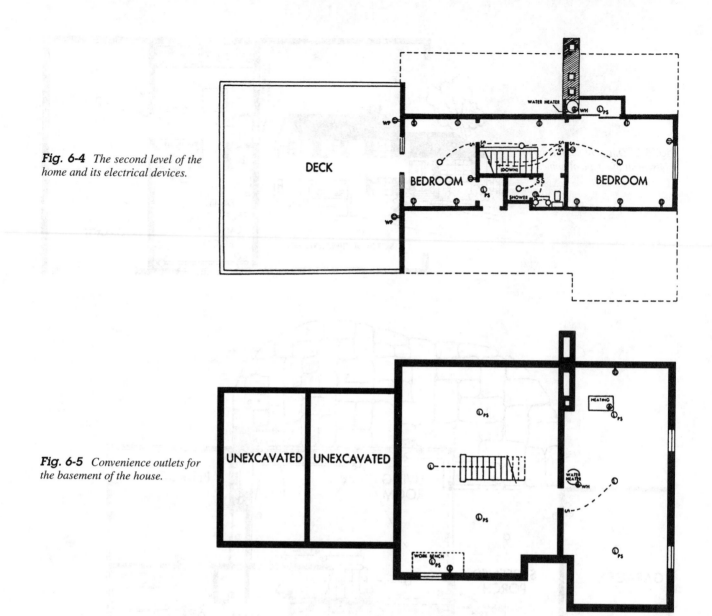

Fig. 6-4 *The second level of the home and its electrical devices.*

Fig. 6-5 *Convenience outlets for the basement of the house.*

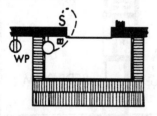

Fig. 6-6 *A weatherproof outlet, preferably near the front entrance, should be located at least 457 millimeters (18 inches) above and be GFCI-protected.*

access to the dwelling unit and to the receptacles. Thus, a GFCI would not be required on a porch or raised portion of a one-family residence. Balconies of apartments and condominiums would not require a GFCI installed in the outside receptacle.

Living room The following planning requirements should be observed.

Lighting provisions Provisions for general illumination of living rooms also apply to sun rooms, enclosed porches, family rooms, television rooms, libraries, dens, and similar living areas (Fig. 6-7). Installation of outlets for decorative lighting accents is recommended, such as picture illumination and bookcase lighting.

One of the requirements for a living room is light. The placement of floor lamps and table lamps is important for reading and enjoying the space. Therefore, it is necessary to provide outlets that will be easily accessible to the lamps.

Receptacles are required on any wall with a width of more than 2 feet. For example, receptacles are not required on a wall with a width of 23 inches. Walls with sliding glass doors or floor-to-ceiling windows must be counted as a wall space that requires receptacles. Here, floor receptacles may need to be installed (Fig. 6-8). From any point along the wall, at

Table 6-1 *Graphical Electrical Symbols for Residential Wiring Plans.*

General Outlets	Switch Outlets
○ Lighting Outlet	S Single-Pole Switch
Ceiling Lighting Outlet for recessed fixture (Outline shows shape of fixture.)	S₃ Three-Way Switch
Continuous Wireway for Fluorescent Lighting on ceiling, in coves, cornices, etc. (Extend rectangle to show length of installation.)	S₄ Four-Way Switch
Ⓛ Lighting Outlet with Lamp Holder	S_D Automatic Door Switch
Ⓛ_PS Lighting Outlet with Lamp Holder and Pull Switch	S_P Switch and Pilot Light
Ⓕ Fan Outlet	S_WP Weatherproof Switch
Ⓙ Junction Box	S₂ Double-Pole Switch
Ⓓ Drop-Cord Equipped Outlet	
Ⓒ Clock Outlet	**Low-Voltage and Remote-Control Switching Systems**
To indicate wall installation of above outlets, place circle near wall and connect with line as shown for clock outlet.	S̲ Switch for Low-Voltage Relay Systems
	MS̲ Master Switch for Low-Voltage Relay Systems
	○_R Relay—Equipped Lighting Outlet
	—·—·— Low-Voltage Relay System Wiring

Convenience Outlets	Auxiliary Systems
Duplex Convenience Outlet	• Push Button
Triplex Convenience Outlet (Substitute other numbers for other variations in number of plug positions.)	Buzzer
Duplex Convenience Outlet — Split Wired	Bell
Weatherproof Convenience Outlet	Combination Bell-Buzzer
Multi-Outlet Assembly (Extend arrows to limits of installation. Use appropriate symbol to indicate type of outlet. Also indicate spacing of outlets as X inches.)	CH Chime
	◇ Annunciator
⊖—S Combination Switch and Convenience Outlet	D Electric Door Opener
⊖—R Combination Radio and Convenience Outlet	M Maid's Signal Plug
⊙ Floor Outlet	Interconnection Box
Range Outlet	T Bell-Ringing Transformer
Special-Purpose Outlet. Use subscript letters to indicate function. DW-Dishwasher, CD-Clothes Dryer, etc.	▶ Outside Telephone
	▷ Interconnecting Telephone
	R Radio Outlet
	TV Television Outlet

Space Heating and Air Conditioning Outlets Use Special-Purpose Outlet and Subscripts	Miscellaneous
▲_SH 230-v Space Heating Outlet	▨ Service Panel
▲_SHF Space Heating Furnace	▬ Distribution Panel
▲_SHB Space Heating Boiler	— — — Switch Leg Indication. Connects outlets with control points.
▲_HP Heat Pump	
▲_CAC Central Air Conditioning	○_a,b ⊖_a,b ▲_a,b □_a,b Special Outlets. Any standard symbol given above may be used with the addition of subscript letters to designate some special variation of standard equipment for a particular architectural plan. When so used, the variation should be explained in the Key of Symbols and, if necessary, in the specifications.
▲_AC 230-v Air Conditioning Outlet	

the floor line, a receptacle must be not more than 6 feet away.

Figure 6-9 shows the location of a receptacle that will allow a lamp to be plugged in with no more than 6 feet from either side of the receptacle.

Extension cords should not be used where people can walk over them.

Electric heaters may be run along the wall and take up floor space when permanently mounted.

Making the receptacle a part of the electric heater can ensure that the extension cord or lamp cord does not fall into the path of hot air.

Fixed room dividers, especially in the living room, must be counted in considering the spacing of receptacles.

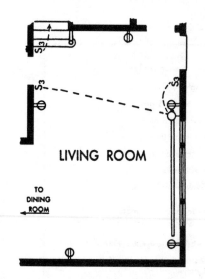

Fig. 6-7 *Some means of general illumination is essential. It may be provided by ceiling or wall fixtures; by lighting in covers, valances, or cornices; or by portable lamps. Provide lighting outlets, wall-switch-controlled, in locations appropriate to the lighting method selected.*

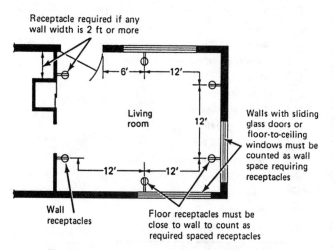

Fig. 6-8 *Location of wall outlets in a living room.*

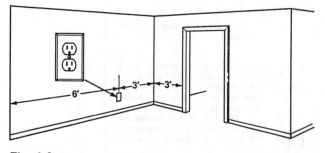

Fig. 6-9 *Location of a receptacle that will permit plugging in of a lamp or appliance.*

Figure 6-10 shows how this is done. Since the two sides of the room divider provide room space, a table lamp may be located where it would not be possible to plug it in without an extension cord. It is possible to place a receptacle on each side of the divider. It is also possible to place a receptacle inside the divider itself.

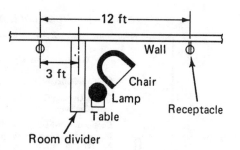

Fig. 6-10 *Fixed wall dividers are counted as wall space when you are figuring the number of outlets needed.*

For switch-controlled outlets, it is recommended that split-receptacle outlets be used. Their use does not limit the location of radios, television sets, clocks, and similar devices. Figure 6-11 shows that one convenience outlet should be installed flush in the mantel shelf if a fireplace is included in the room.

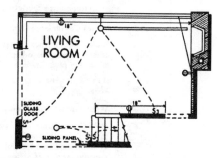

Fig. 6-11 *Location of convenience outlets in a living room.*

Special-purpose outlets It is recommended that one outlet for a room air conditioner and outlets for space heating be installed if central systems are not planned.

Dining areas The following planning requirements should be observed.

Lighting provisions Each dining room, or dining area combined with another room, or breakfast nook shall have at least one lighting outlet, wall-switch-controlled (Fig. 6-12). When a dining or breakfast table

Fig. 6-12 *Lighting outlets for dining areas are normally located over the probable location of the dining or breakfast table, to provide direct illumination.*

is to be placed at the table location, the switch should be placed just above table height. Built-in counter space should have an outlet provided above counter height for portable appliances.

Convenience outlets   These should be of the split-receptacle type, for connection to small appliances.

Special outlets   It is recommended that outlets for space heating and one outlet for a room air conditioner be installed whenever central systems are not planned.

Bedrooms   The following planning requirements should be observed.

Lighting provisions   Light fixtures over full-length mirrors, or a light source at the ceiling located in the bedroom and directly in front of the clothes closets, may serve as general illumination (Fig. 6-13).

Master-switch control in the master bedroom, as well as at other strategic points in the home, is suggested for selected interior and exterior lights.

Convenience outlets   It is recommended that convenience outlets be placed only 3 to 4 feet from the centerline of the probable bed locations. The popularity of bedside radios and clocks, lamps, and electric bed covers makes increased plug-in positions at bed locations essential. Triplex or quadruplex convenience outlets are therefore recommended at these locations (Fig. 6-14).

Furthermore, at one of the switch locations, there should be a receptacle outlet for a vacuum cleaner, floor polisher, or other portable appliances (Figs. 6-15 and 6-16).

Special-purpose outlets   It is recommended that outlets for space heating and one outlet for a room air conditioner be installed in each bedroom if central systems are not provided.

Bathroom and lavatories   The following planning requirements should be observed.

Lighting provisions   A ceiling outlet located in line with the front edge of the wash basin will provide lighting for the mirror, general room lighting, and safety lighting for a combination shower and tub (Figs. 6-17 and 6-18).

When more than one mirror location is planned, equal consideration should of course be given to the

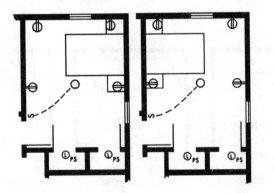

Fig. 6-13 *Good general illumination is particularly important in a bedroom. This can be provided from a ceiling fixture or from lighting valances, covers, or cornices. Provide the proper wall switch controls.*

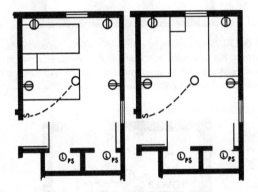

Fig. 6-14 *Start with an imaginary centerline through each probable bed location. Install outlets on each side of the centerline and within 6 feet of it. No point along the floor line in any other usable wall space should be more than 6 feet from an outlet on that wall. Add outlets as needed to achieve this.*

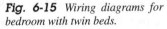

Fig. 6-15 *Wiring diagrams for bedroom with twin beds.*

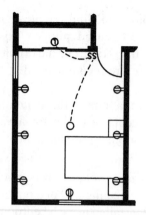

Fig. 6-16 *The same bedroom with wiring adaptation to double-bed arrangements.*

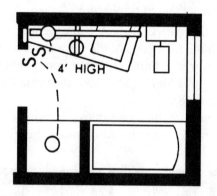

Fig. 6-17 *Illumination of both sides of the face at a mirror is important. Keep in mind that a single concentrated light, on either the ceiling or sidewall, is usually not acceptable. All lighting outlets should be wall-switch-controlled.*

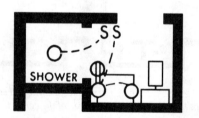

Fig. 6-18 *One outlet near the mirror, 3 to 5 feet above the floor, is required. See Code requirements about GFCI and outlets.*

lighting in each case. A switch-controlled night-light may be installed where desired.

If an enclosed shower stall is planned, an outlet for a vapor-proof luminaire is sometimes installed, controlled by a wall switch outside the stall.

Convenience outlets   It is suggested that an outlet be installed at each separate mirror or vanity space. An outlet should also be installed at any space that might accommodate an electric towel dryer, electric razors, or electric toothbrushes. A receptacle that is part of a bathroom lighting fixture should not be considered as satisfying this requirement for small appliances, unless

it is rated at 15 amperes and wired with at least 15-ampere-rated wires.

Code requirements   Any receptacle that is an integral part of an appliance, cabinet, or lighting fixture may not be used to satisfy the specific receptacle requirements for bathrooms. Every bathroom must have at least one wall-switch controlled lighting outlet. This does not mean a pull chain or a canopy will serve as the requirement.

Take a close look at Fig. 6-19 to see what is and what is not considered a bathroom according to Code requirements. A receptacle is not usually required in the area with the tub and toilet. However, if one is installed in these areas, it must be protected by a GFCI.

Figure 6-20 shows a typical bedroom suite in a one-family house or apartment. Note the location of the

Fig. 6-19 *What is and what is not a bathroom as far as the Code is concerned.*

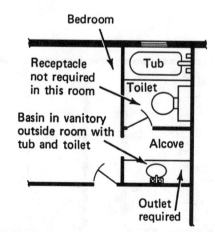

Fig. 6-20 *A typical bedroom suite with a bathroom off the bedroom. Note the location of the receptacle.*

bathroom off the bedroom. Even with the basin in the vanity outside the tub or toilet area it does require a receptacle at the basin. A receptacle is not required in the room where the tub is located. Figure 6-21 indicates possible receptacle locations.

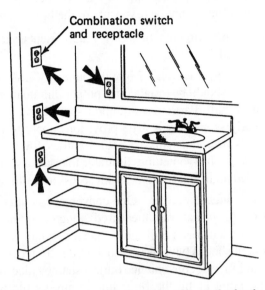

Fig. 6-21 *These are possible locations of the outlet for the bathroom. The location will vary, depending on the wall space available. A receptacle in the medicine cabinet or in the bathroom light does not count as the required receptacle.*

Kitchen The following planning requirements should be observed.

Lighting provisions Provide outlets for general illumination and for lighting at the sink. These lighting outlets should be wall-switch-controlled. Lighting design should provide for illumination of work areas, sink, range, counters, and tables. Under-cabinet lighting fixtures, within easy reach, may have local switch control. In some kitchens, consideration should also be given to outlets to provide inside lighting of cabinets.

One outlet for the refrigerator is needed; and one outlet for each 4 linear feet of work surface frontage, with at least one outlet to serve each work surface. If a planning desk is to be installed, one outlet is needed for this area. (Work surface outlets are to be located about 44 inches above the floor line.) Table space should have one outlet, preferably just above the table level (Fig. 6-22).

Convenience outlets An outlet is recommended at any wall space that may be used for an iron or for an electric toaster. Convenience outlets in the kitchen should be of the split-receptacle type, for connection to small appliances.

Special-purpose outlets One type of appliance that is helpful in the kitchen is a clock. It should be in

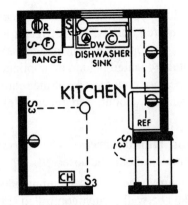

Fig. 6-22 *Wiring diagram for a kitchen.*

a location easily visible from all parts of the kitchen. It requires a recessed receptacle, with a clock hanger (Fig. 6-23). An outlet for a freezer, either in the kitchen or in some other convenient location, is often necessary.

Code requirements Receptacles are required at each counter space that is wider than 12 inches. If it is located inside the dining room, the counter must also have a receptacle if more than 12 inches of counter space is usable (Fig. 6-24).

Some receptacles are rendered inaccessible by the installation of appliances that are either fastened in place or positioned in a space that is dedicated or assigned

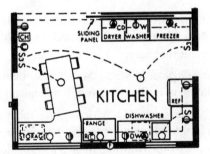

Fig. 6-23 *The following outlets are usually necessary: one for a freestanding range, or for a built-in range surface unit, and one for each built-in range oven (if each oven is a separate unit); one each for a ventilating fan, dishwasher, and food waste disposer (if necessary plumbing facilities are installed); and one for an electric clock.*

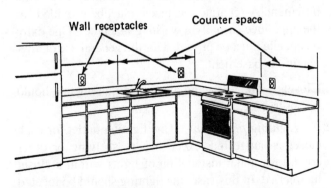

Fig. 6-24 *Required spacing for receptacles in a kitchen.*

permanently to an appliance. These receptacles are not counted as required outlets. The refrigerator would be an example of an appliance dedicated to a space.

A receptacle located behind an appliance and not accessible does not count as one of the required countertop receptacles. Countertop receptacles must be accessible.

Laundry and laundry areas The following planning requirements should be observed.

Lighting provisions It is recommended that all laundry outlets be wall-switch-controlled (Fig. 6-25).

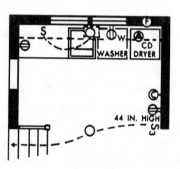

Fig. 6-25 *For a complete laundry, lighting outlets are installed over the work areas such as laundry tubs, sorting table, washing, ironing, and drying centers. For laundry trays in an unfinished basement, one ceiling outlet is required centered over the trays. One convenience outlet, or more if desired, shall be installed.*

Convenience outlets Convenience outlets in the laundry area should be connected to a 20-ampere branch circuit that serves no other areas.

Special-purpose outlets Sometimes, plans require the installation of outlets for a ventilating fan and a clock.

Code requirements At least one receptacle—it may be a single, duplex, or triplex—must be installed for the laundry of a dwelling. Such a receptacle and any other receptacles for special appliances must be placed within 6 feet of the intended location of the appliance. The Code requires a receptacle outlet in a basement in addition to any receptacle outlet that may be provided as the required receptacle to serve a laundry area in the basement. A separate receptacle must be provided for the appliance (usually a washing machine). One extra receptacle must be provided for the general use of anyone in the basement.

Closets The following planning requirements should be observed.

Lighting provisions One lighting outlet for each closet is suggested, except where shelving or other conditions make installation of lights within a closet ineffective. In this case, the lighting should be located in the adjoining space to provide light within the closet.

Installation of wall switches near a closet door, or door-type switches, is recommended.

The light fixture must be so located that the bulb or fixture will not be within 12 inches of any shelf in the closet. Nothing piled up on the shelf should be within 12 inches of the lighting fixture. This is done to prevent accidental fires from overexposure to the light source.

Halls The following planning requirements should be observed.

Lighting provisions Lighting outlets, wall-switch controlled, should be installed for proper illumination of the entire hall. Particular attention should be paid to irregularly shaped areas. The provisions apply to passage halls, reception halls, vestibules, entries, foyers, and similar areas. It is sometimes desirable to install a switch-controlled night-light in a hall with access to bedrooms.

Convenience outlets One convenience outlet for each 15 feet of hallway, measured along centerline, is recommended. Each hall more than 25 square feet in floor area should have at least one outlet. In reception halls and foyers, convenience outlets shall be placed so that no point along the floor line in any usable wall space is more than 10 feet from an outlet in that space.

It is further recommended that at one of the switch outlets in the hallway, a convenience receptacle be provided for connection of a vacuum cleaner or other floor-cleaning device.

Code requirements In hallways, stairways, and at outdoor entrances, remote, central, or automatic control for lighting shall be permitted according to the Code. This is an alternative to the wall-switch-controlled light.

Stairways The following planning requirements should be observed.

Lighting provisions Wall or ceiling outlets should provide adequate illumination of each stair flight. Outlets shall have multiple-switch control at the head and foot of the stairway, so arranged that full illumination may be turned on from either floor, but that lights in halls furnishing access to bedrooms may be extinguished without interfering with ground-floor usage.

These provisions are intended to apply to a stairway that connects finished rooms at either end.

Whenever possible, switches should be grouped together. They should never be located so close to steps that a fall might result while one is reaching for a switch.

At intermediate landings of a large stairway (depending on planned usage), an outlet is often recommended for a decorative lamp, night-light, or cleaning equipment.

Recreation, TV, or family rooms The following planning requirements should be observed.

Lighting provisions   Selection of a lighting method for these rooms should take into account the types of activities for which the room is planned (Fig. 6-26). General illumination is essential in a family, recreation, play, or TV room. It may be provided by ceiling or wall fixtures, or by lighting in coves, valances, or cornices. Provide lighting outlets, wall-switch controlled, in appropriate locations. Convenience outlets should be placed so that no point along the floor line in any usable wall space is more than 6 feet from an outlet.

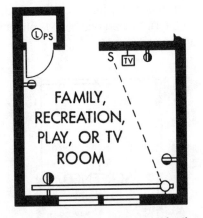

Fig. 6-26 *Lighting methods for recreation or family rooms should be selected according to the activities for which the room is planned.*

Convenience outlets   Convenience outlets in family rooms should be the split-receptacle type for connection to small appliance circuits.

If any of these rooms contains a fireplace, one convenience outlet should be installed flush in the mantle shelf. Outlets for the use of a clock, radio, television, ventilating fan, DVD, VCR, and the like should be located in relation to their intended use.

Special-purpose outlets   Outlets for space heating and one outlet for a room air conditioner should be installed if central systems are not planned.

Utility rooms or space   The following planning requirements should be observed.

Lighting provisions   Lighting outlets in a utility room (or space) should be placed to illuminate the furnace area, and work bench, if planned. At least one lighting outlet is to be wall-switch-controlled (Fig. 6-27). One convenience outlet, preferably near the furnace, or near any planned workshop, should be installed. Special-purpose outlets, such as one each for the water heater, boiler, furnace, or other equipment, should be located convenient to the appliance or equipment.

Convenience outlets   The washing machine outlet should be connected to a 20-ampere, laundry-area branch circuit.

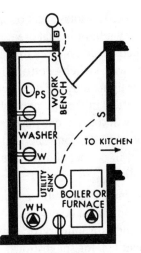

Fig. 6-27 *A wall-switch-controlled outlet is usually necessary to control the light in a utility room upon entering, because often this room has no windows.*

Basements   The following planning requirements should be observed.

Lighting provisions   In basements with finished rooms, with garage space, or with other direct access to outdoors, the stairway lighting provisions stated previously apply. For basements that will be infrequently visited, a pilot light should be installed in conjunction with the switch at the head of the stairs.

Convenience outlets   Basement lighting outlets should illuminate designated work areas or equipment locations, such as furnace, pump, and workbench (Fig. 6-28). Additional outlets should be installed near the foot of the stairway, in each closed space, and in open

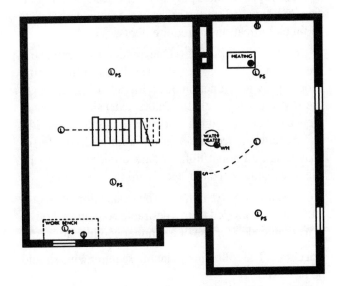

Fig. 6-28 *Basement convenience outlets are useful near the furnace; at the play areas; for basement laundries, darkrooms, and hobby areas; and for appliances such as dehumidifiers and portable space heaters.*

spaces so that each 150 square feet of open space is adequately served by a light in that area.

In unfinished basements, the light at the foot of the stairs should be wall-switch-controlled near the head of the stairs. Other lights may use a pull chain for control. Convenience outlets in the basement should number at least two. For a workbench, one outlet should be placed at this location. In addition, one outlet for electrical equipment used in connection with furnace operation should be installed.

Special-purpose-outlet If a freezer is located here, an outlet for it must be installed. You may want to use the basement as a laundry or utility room. If so, check these headings for suggestions.

Code requirements At least one lighting outlet must be installed in every attic, underfloor space, utility room, and basement if it is used for storage or if it contains equipment requiring servicing. The lighting fixture does not necessarily have to be wall-switch-controlled. A pull chain fixture is acceptable.

Accessible attic The following planning requirements should be observed.

Lighting provisions One outlet for general illumination and a wall switch controlled from the foot of the stairs are the minimum. When no permanent stairs are installed, this lighting outlet may be pull-chain-controlled, if located over the access door. Where an unfinished attic is planned for later development into rooms, the attic lighting outlet shall be switch-controlled at the top and bottom of the stairs. One outlet for each enclosed space is recommended.

These provisions apply to unfinished attics. For attics with finished rooms or spaces, see the appropriate room classification for requirements.

Convenience outlets One outlet for general use is the minimum. If an open stairway leads to future attic rooms, provide a junction box with direct connection to the distribution panel, for future extension to convenience outlets and lights when the rooms are finished.

A convenience outlet in the attic is desirable for providing additional light in dark corners and for the use of a vacuum cleaner and cleaning accessories.

Special-purpose outlet The installation of an outlet, switch-controlled from a desirable point in the house, is recommended for use of a summer cooling fan.

Porches The following planning requirements should be observed.

Convenience outlets Each porch, breezeway, or other similar roofed area of more than 75 square feet in

floor area should have a lighting outlet, wall-switched-controlled. Large or irregularly shaped areas may require two or more lighting outlets. One convenience outlet (weatherproof, if exposed to moisture) for each 15 feet of wall that borders a porch or breezeway is recommended (Fig. 6-29). Multiple-switch control should be installed at entrances if the porch is used as a passage between house and garage.

The split-receptacle convenience outlet is intended to be connected to a three-wired appliance branch circuit. If this area is considered an outdoor dining area, all outside outlets must have a ground-fault circuit interrupter (GFCI). At least one outside outlet is required for any newly constructed house.

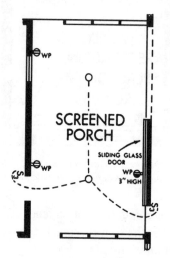

Fig. 6-29 *One outlet, weatherproof if exposed to moisture, for each 15 feet of wall bordering the porch or breezeway is recommended. If the outlet is accessible by someone standing on the earth, then it must also be GFCI-protected. One or more such outlets should be controlled by a wall switch inside the door.*

Terraces and patios The following planning requirements should be observed.

Lighting provisions One or more outlets on the building wall, or other convenient location in the area to provide fixed illumination, are recommended. Wall-switch control shall be provided inside the house.

Convenience outlets One weatherproof outlet located at least 18 inches above the grade line is needed for each 15 linear feet of house wall bordering terrace or patio. It is recommended that one or more of these outlets be wall-switch-controlled from inside the garage.

Garages or carports If a garage is to be used for additional purposes, such as a work area, storage closets, laundry, attached porch, or the like, the rules apply that are appropriate to those uses (Fig. 6-30).

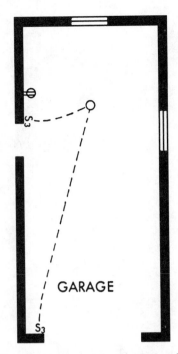

Fig. 6-30 *The garage or carport should contain at least one ceiling outlet, wall-switch-controlled for a one- or two-car garage or storage area. If the garage has no covered access from the house, provide one exterior outlet, multiple-switch- controlled, from both garage and residence. One plug-in outlet for a one- to two-car storage plug-in outlet for an area is necessary. The outlet should be GFCI-protected.*

An exterior outlet, wall-switch controlled, is recommended for all garages. Additional exterior outlets are often desirable, even if no specific additional use is planned for the garage. Long driveways warrant additional illumination, such as a post light, wall-switch-controlled from inside the house.

If a freezer or refrigerator, workbench, automatic door opener, or car heater is planned for the garage, outlets for these uses should be provided.

Code requirements   A vehicle door in an attached garage is not considered as an outdoor entrance. This means the Code does not require a light outlet at the garage door. It is provided with light by the car's headlights.

If it is an attached garage, there must be at least one receptacle. It should have GFCI protection. GFCI protection is required for all 120-volt, single-phase, 15- and 20-ampere receptacles installed in a garage. This is so because homeowners use outdoor appliances plugged into the garage outlet. This gives protection for lawnmowers, snowblowers, hedge trimmers, and hand-held electric drills. A carport attached to the house should have a GFCI-protected outlet.

Exterior grounds   In addition to terrace and patio lighting, outlets under the eaves on the exterior of a house and garage are sometimes recommended for protective or decorative lighting. Switch control should be from within the house. Multiple- and master-switch control is desirable also.

Weatherproof floodlights and spotlights are available to show the beauty of a garden by night. These are designed for use with ether 115-volt lamps or lamps at reduced voltage supplied through safety transformers. Fixtures may be permanently wired and switched from inside the house, or they may be served from outdoor weatherproof convenience outlets. Lighting for sports (such as tennis, volleyball, or badminton) deserves consideration where the grounds are suitable for this purpose.

Underwater and area lighting for a swimming pool and the power requirements for a pool circulating pump and filter require careful design and installation by specialists in this field.

In general, it is recommended that all outdoor lighting outlets be switch-controlled from within the house. In addition, certain ones, such as post lights and protective lighting, lend themselves to photoelectric cell or time switch control.

In climates where buildup of snow and ice on roofs is a problem, a "snow-melting" (heating) cable may be installed at the lower edge of the roof and in the gutters and downspouts. This provides excellent protection at low cost against water leakage into ceilings and outside walls of living quarters.

Motor-operated valves for buried lawn sprinkler systems are available. These can be arranged for manual or automatic operation. However, the water-operated ones are quieter.

It is best to check the architect's specifications for outlets with the proper person to make sure the proper number and size are installed. The best way to check is to ask the electrician if the wiring and installation are inspected by a local electrical inspector. In some cases, the insurance company will not insure a house unless it has been inspected and certified to meet the local codes. In some cases the National Electrical Code is specified as the bare minimum. The local code adds to this to meet special regional requirements.

It is very expensive to change plans once construction begins. The proper thing to do is to go over the plans with a qualified electrician. Then tell the electrician your problems and what you want to do with each room before construction starts. This way the proper electrical system can be designed and installed.

To be realistic, no house is ever properly wired. The persons living in a house change in their requirements for electrical power. They buy new devices and they rearrange the furniture. In some instances, use of an extension cord can solve a problem. However, the use of extension cords should be limited. The proper wire size should always be used for a particular circuit's requirements.

As an electrician, you should check the latest issue of the National Electric Code for any new procedures or limitations. Keep current.

7
CHAPTER

Equipment and Tools for Wiring

FOR AN ELECTRICIAN TO DO THE JOB PROPERLY, it is necessary to have the right tools. Tool requirements vary with the job. Some need only a minimum of tools. Others require some rather complicated devices.

The more complicated tools are used by the electrician who works on commercial and industrial installations. Some of these tools, which are very expensive, are bought by the electrical contractor. It is necessary, however, for an electrician to have used such tools, or at least to have knowledge of their existence.

This chapter will endeavor to show both the simple and the complicated tools that are used every day by an electrician. Some electricians have had to work with devices they improvised themselves. These have later become standard equipment for certain jobs. The ingenuity of electricians is well known to those people associated with them. Their ability to get a job done efficiently is also well known. Yet, most of this efficiency is due to the particular individual's skill in utilizing the equipment at hand.

BASIC TOOLS

Figure 7-1 illustrates the minimum tools necessary for house wiring or for adding an outlet or switch to existing wiring.

• *Hammer.* Although primarily used for driving nails and staples or for fastening hangers, a hammer may be used to anchor an object to concrete (Fig. 7-2). In Fig. 7-3 a hammer is being used to attach a box to a 2-inch × 4-inch stud. The hammer has many uses to an electrician.

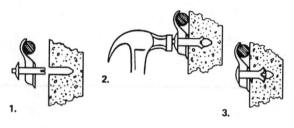

Fig. 7-2 *Using a hammer to set an anchor.*

• *Bit.* A drill bit may be used to drill holes in wood or soft metal. Some more extensive uses will be shown later in this chapter. The bit in Fig. 7-1 has been designed for use in a brace, for operation by hand.

• *Bit brace.* This tool is very useful in drilling holes for the location of boxes and wires, and for fishing wiring out of hard-to-reach locations (Fig. 7-1).

• *Keyhole saw.* The keyhole saw cuts circles and irregular shapes (Fig. 7-1).

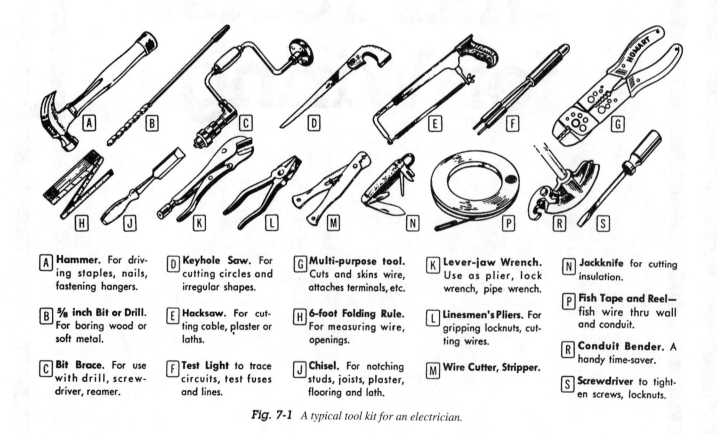

A	**Hammer.** For driving staples, nails, fastening hangers.
B	**⅝ inch Bit or Drill.** For boring wood or soft metal.
C	**Bit Brace.** For use with drill, screwdriver, reamer.
D	**Keyhole Saw.** For cutting circles and irregular shapes.
E	**Hacksaw.** For cutting cable, plaster or laths.
F	**Test Light** to trace circuits, test fuses and lines.
G	**Multi-purpose tool.** Cuts and skins wire, attaches terminals, etc.
H	**6-foot Folding Rule.** For measuring wire, openings.
J	**Chisel.** For notching studs, joists, plaster, flooring and lath.
K	**Lever-jaw Wrench.** Use as plier, lock wrench, pipe wrench.
L	**Linesmen's Pliers.** For gripping locknuts, cutting wires.
M	**Wire Cutter, Stripper.**
N	**Jackknife** for cutting insulation.
P	**Fish Tape and Reel—** fish wire thru wall and conduit.
R	**Conduit Bender.** A handy time-saver.
S	**Screwdriver** to tighten screws, locknuts.

Fig. 7-1 *A typical tool kit for an electrician.*

Fig. 7-3 *Nailing a box to a stud. Some are made of plastic and others of metal, but both come with nails inserted in the box.*

- *Hacksaw.* Used for cutting cable (BX or Romex), the hacksaw can cut plaster or laths, or any metallic conduit, thin-wall or rigid (Fig. 7-1).
- *Test light.* A test light is used to trace circuits, test fuses, and test circuits to determine whether power has been applied (Figs. 7-1 and 7-4). It can also indicate the voltage and whether it is AC or DC.

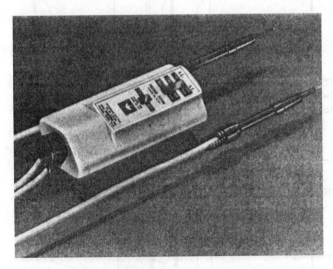

Fig. 7-4 *A circuit tester.*

- *Multipurpose tool.* Used to strip wires and crimp connectors, this tool will also cut screws without ruining the threads (Fig. 7-1).

- *Folding rule.* This standard 6-foot folding rule is in every electrician's tool kit. It is used to check the proper location of boxes, wiring, and figures (Fig. 7-1).
- *Chisel.* A necessary tool for making the required small changes in the woodwork when one is mounting a fixture or box, a chisel may also be used to notch studs, joists, plaster, flooring, and lath (Fig. 7-1).
- *Lever-jaw wrench.* This vise grip can be used as a pliers, lock wrench, or pipe wrench (Fig. 7-1).
- *Side cutters (sometimes called linesmen's pliers).* This tool probably gets as much use as any other in the electrician's tool kit. It can be used for gripping locknuts, cutting wires, and breaking fins off dimmers (Fig. 7-5). It can be used to work on floor boxes. This tool also serves as a crimper of new-type connectors for Romex (Fig. 7-6) and for removing twist-outs and knockouts.
- *Wire cutter, stripper.* A useful device for stripping wire, it is adjustable for the wire size being used. This

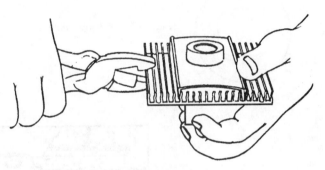

Fig. 7-5 *Pliers used to break off the fins on a light dimmer.*

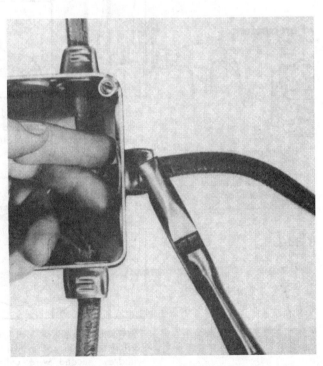

Fig. 7-6 *Pliers are being used to twist Romex connector.*

tool makes it possible to strip insulation without cutting or marring the copper or aluminum conductor. It may be used to cut wires (Fig. 7-1).

- *Jackknife.* A jackknife has many uses. The electrician cuts insulation with it without scoring the conductor. In addition, it might be used to cut tape (Fig. 7-1).

- *Fish tape and reel.* This device, as the name suggests, is used to fish wire through walls and conduit (Figs. 7-7 and 7-8).

- *Conduit bender.* The conduit bender is handy for making bends needed for turns in the EMT (thin-wall conduit) or rigid conduit. Some special tools have been designed and manufactured for the electrician working with both rigid and plastic conduit (Fig. 7-1).

- *Romex stripper, or ripper.* The Romex stripper, or ripper (Fig. 7-9), is used to remove the thick coating of insulation from the two or three conductors in the cable. The ripper has a tip that sticks up inside the unit and will fit into the Romex insulation when the tool is squeezed. The tool is pulled to rip the outside coating of insulation. Then the insulation is cut with a pair of wire cutters, or with a diagonal cutter such as the one in Fig. 7-10.

Fig. 7-9 *A Romex stripper, or ripper.*

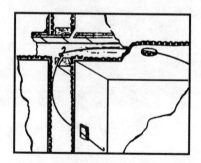

Fig. 7-10 *A wire cutter.*

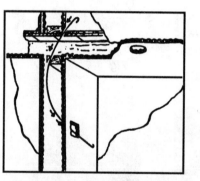

Fig. 7-7 *A fish tape and reel.*

Fig. 7-8 *Using a fish wire.*

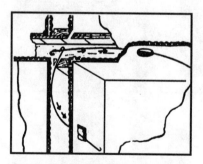

Push 12-foot fish wire, hooked at two ends, through hole on 2nd floor. Pull one end out at switch outlet on 1st floor.

Then withdraw either wire (arrows) until it hooks the other wire; then withdraw second wire until both hooks hook together.

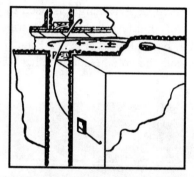

Next, push 20-25-foot fish wire, hooked at both ends thru ceiling outlet (arrows). Now fish until you touch the first wire.

Lastly, pull shorter wire thru switch outlet. When hook from long wire appears, attach cable and pull thru wall and ceiling.

- *Standard wire gage.* This gage checks the size of wire (Fig. 7-11). By placing a wire through the slot (not the hole) in the gage, it is possible to determine the size of the wire (Fig. 7-12).

Fig. 7-13 *A handsaw.*

Fig. 7-14 *An electrician drilling a hole in a stud with a battery-operated electric drill.*

Fig. 7-11 *A wire gage.* (Brown & Sharpe)

Fig. 7-12 *Using a wire gage. Note: Wire thickness is measured by the slot, not the hole.* (Brown & Sharpe)

- *Handsaw.* A handsaw is handy for cutting studs or for making a box fit a bracket or fixture (Fig. 7-13).
- *Electric drill.* Electric drills have, in many cases, replaced the hand brace and bit. The drill shown in Fig. 7-14 is battery-operated. It is very handy on new construction sites where power is not yet available.

- *Hole saw.* A number of hole saws are available. The saw in Fig. 7-15 first drills a pilot hole, then cuts a hole the diameter of the curved saw blade. In Fig. 7-16, you can see the different diameters available (only one saw blade is left in the tool when it is used).

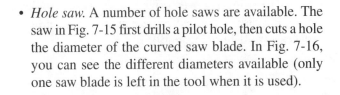

Fig. 7-15 *The parts of a hole saw.*

Fig. 7-16 *Hole saw blades. You take out the ones not being used.*

- *Spin-tite wrenches and screwdrivers.* Spin-tite wrenches are screwdrivers with sockets on the end. Figures 7-17 and 7-18 show the variety of screwdrivers usually included in an electrician's tool kit. Both the standard-type head and the Phillips head are included.

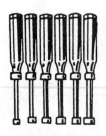

Fig. 7-17 Spin-tite socket wrenches.

Fig. 7-18 Screwdrivers.

- *Chalk line.* This gadget comes in handy when a straight line is needed for a distance of more than a few feet. The string is covered with chalk as it is pulled from the case. When pulled tight and snapped, it produces a white line (Fig. 7-19).

Fig. 7-19 A chalk line.

- *Fuse puller.* A plastic-handled fuse puller is a tool used to safely extract fuses from industrial or commercial distribution boxes (Fig. 7-20).

Fig. 7-20 A fuse puller. This is used for cartridge-type fuses.

- *Awl.* The awl has a variety of uses. It can be used to mark lines or make holes, and as a tool for prying staples loose (Fig. 7-21).

Fig. 7-21 An awl. It is used to start screws by making a pilot hole or for any number of other applications.

- *Soldering irons and torches.* Soldering irons and guns are common tools for electronics technicians, and electricians also use them on occasion. However, if there is no electrical power available at a construction site, it becomes necessary to find tools that operate on other power. A propane torch is efficient here. It may be used for soldering either small or large jobs (Fig. 7-22).

Fig. 7-22 A propane torch.
(Bernzomatic)

There are other tools that the electrician can obtain for use on special jobs. Those listed here represent the necessities. A toolholder, shown in Fig. 7-23, comes in handy.

The offset needed to match up knockout holes in boxes to a piece of flat-mounted conduit can be accomplished with a special bender for $\frac{1}{2}$-inch and $\frac{3}{4}$-inch thin-wall EMT (Figs. 7-24 and 7-25). This bend in the conduit eliminates the need for connectors whenever the EMT is used in exposed locations, thus preventing a lot of trial-and-error bending and cutting (Fig. 7-26). Tools are available for all special jobs.

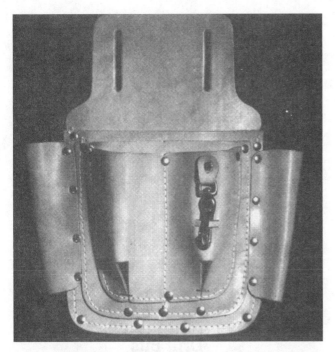

Fig. 7-23 *An electrician's toolholder.* (Eller)

Fig. 7-25 *A hand bender for EMT and rigid conduit. Note the guide marks on the head of the bender.* (Greenlee)

Fig. 7-24 *Using a conduit bender to make a fit.* (Greenlee)

Fig. 7-26 *A utility box surface-mounted with a offset bend.* (Greenlee)

Fig. 7-27 *A hacksaw.* (Greenlee)

Saws and Chisels

Saws   One of the most-used saws on the job is the *hacksaw* (Fig. 7-27). It comes in handy for cutting pipe, conduit, or any metal. A *compass saw* is used for cutting wood to make way for conduit, or a box, or cabinet. Figure 7-28 shows a blade for a compass saw. The saw handle is similar to that of a hacksaw and shaped to fit the hand.

Fig. 7-28 *A compass saw blade for woodwork.* (Greenlee)

The *flexsaw* is a compact saw for close-quarter jobs (Fig. 7-29). It uses a hacksaw blade, which flexes for a cut flush to the surface. It is used to cut metal.

Fig. 7-29 A flexsaw. (Greenlee)

Hacksaw blades suitable for the type of material to be cut should be chosen. They are available in 10-inch and 12-inch lengths. But the number of teeth per inch should depend on the shape and size of material being cut. The three teeth designations available are 18, 24, and 32 teeth per inch. The more teeth the better for thin material, or for small-diameter pipe.

Chisels Chisels are used to remove wood from studs or framing to allow clearance for mounting conduit or metal boxes. There are a number of chisels available for electricians or carpenters. Here are some usually chosen by electricians or line workers.

The *utility chisel* in Fig. 7-30 is a special-purpose type used by utility and telephone line workers to cut daps in poles for cross-arms. The blade is 2 inches wide and heat-treated for hard wear and use.

Fig. 7-30 A utility chisel. (Greenlee)

The *firmer chisel,* with beveled edges, is made of high-carbon tool steel (Fig. 7-31).

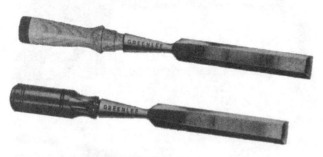

Fig. 7-31 Firmer chisels. (Greenlee)

The *framing chisel* can be used for heavy framing, or as a wrecking tool, or like the firmer chisel, to chip wood. Notice that it does not have beveled edges. This is one way to distinguish it from the utility chisel (Fig. 7-32).

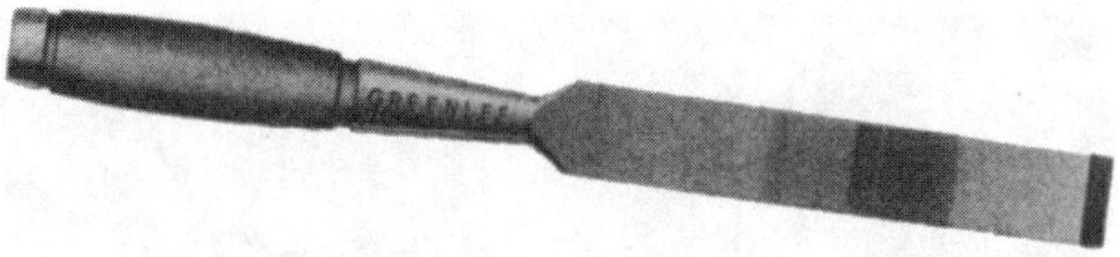

Fig. 7-32 A framing chisel. (Greenlee)

The *gouge* is a woodworker's tool. The electrician uses it to remove excess wood for the installation of conduit (Fig. 7-33).

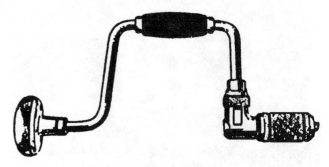

Fig. 7-33 A wood gouge. (Greenlee)

Drill Bits

Woodworking tools are valuable to an electrician who must makes holes in wood to pull wire or place conduit. A number of drills are available for boring holes in wood. A hand-operated drill, called a *brace,* takes a number of drill bits suitable for each job (Fig. 7-34).

Fig. 7-34 A brace without bit.

The bit brace uses the principle of the wheel and axle in creating a driving force. The jaws of the chuck hold the bit in the brace during operation. Pressure is applied to the *head* of the brace (the rounded, wooden knob on the end). A turning motion is applied by grasping the *handle* in the middle of the brace and rotating it. Drill bits have a screw tip to catch the wood and pull the cutting edge forward. The *twist* of the bit causes a smooth flow of wood chips to the outside of the hole.

Some braces have a ratchet wheel that allows for turning the bit in a clockwise direction only, but allows the *bow* (bent part of the brace) to be turned in the opposite direction while the bit is stationary in the wood. It is then possible to bore a hole in a piece of wood where the full swing of the bow is not possible. A *cam ring* can be turned to allow for ratchet action or for full operation.

An electrician should be able to identify the proper drill bit for a job. Some bits are more efficient than others due to their design and availability in a number of sizes. The sizes of the following bits are given in customary measurement. Drill bits are available in metric sizes. Eventually, such metric drill bits will be more commonly used.

Machine spur bit The bit in Fig. 7-35 is properly referred to as a *machine spur bit.* It is a flat, wood-boring power bit designed for use in electric drills, to bore holes in any wood, at any angle, and at any speed. It is usually a little over 6 inches in length and comes in cutter sizes ranging from $1/4$ inch to $1 1/2$ inches, in increments of $1/16$ inch.

Fig. 7-35 *A machine spur bit.* (Greenlee)

Multispur pipe bit It has three milled flats for use in portable electric or pneumatic drills, and in stationary boring machines (Fig. 7-36).

Fig. 7-36 *A multispur pipe bit.* (Greenlee)

Plain-type expansive bits These are used in a brace and have a clamp and screw for locking the cutter in place (Fig. 7-37). This bit may be obtained in $1/2$- to $1 1/2$-inch, $7/8$- to $1 3/4$-inch, and $1 3/4$- to 3-inch cutters.

Fig. 7-37 *An expansive bit.* (Greenlee)

Auger bits A *bellhanger's drill* is used for telephone and other wiring installations. Sizes are $3/16$ to $3/4$ inch in

$1/16$-inch increments. It comes in 12-, 18-, and 24-inch lengths (Fig. 7-38).

Fig. 7-38 *A bellhanger's drill bit.* (Greenlee)

Unispur electrician's power bit This is a double-twist, smooth-boring bit used by electricians, plumbers, and pipe fitters (Fig. 7-39). It is especially designed for use in electric drills. The screw point feeds at 16 turns to the inch. Note the two setscrew holes for use in a power bit extension or brace adapter. The overall length is $8 1/2$ inches. It comes in $5/8$- to $1 1/4$-inch size with $1/16$- and $1/8$-inch intervals, depending upon the diameter needed.

Fig. 7-39 *A unispur power bit.* (Greenlee)

Single-twist power bit This is a smooth-boring bit designed for electric drills (Fig. 7-40). It has excellent chip clearance due to its single-twist design. It may be used to bore holes from $5/8$ to 1 inch with $1/16$-inch difference between various drill sizes. Overall length is from 5 to 7 inches, depending upon the drill size. With extensions, it will drill through floor joists, studs, or existing construction for installation of Romex.

Fig. 7-40 *A single-twist power bit.* (Greenlee)

Solid-center power-car bit It has a solid center to add stiffness and strength (Fig. 7-41). It has a single twist for excellent chip removal from the cutting area. The screw point is of a medium pitch. It is designed for use in three-jaw chucks of $3/8$ inch and larger. It can be used with a bit extension. Overall length is about 16 inches, available in sizes from $3/8$ to 1 inch in sixteenths of an inch.

Fig. 7-41 *A solid-center power-car bit.* (Greenlee)

Power-pipe bit It is ideal for boring holes in wood to allow passage of pipe, conduit, and tubing, and for other uses requiring large holes (Fig. 7-42). It is designed for

Fig. 7-42 *A power-pipe bit.* (Greenlee)

rough, rugged boring. Lengths run from 14 to 17 inches overall, with cutter sizes available from $1\frac{1}{4}$-inch diameter for $\frac{3}{4}$-inch pipe, $1\frac{1}{2}$-inch diameter for 1-inch pipe, $1\frac{3}{4}$-inch diameter for $1\frac{1}{4}$-inch pipe, 2-inch diameter for $1\frac{1}{2}$-inch pipe, and $2\frac{1}{2}$-inch diameter for 2-inch pipe.

Impact-wrench bit   This can be identified by the grooved portion near the end of the bit (Fig. 7-43). This single-spur (spur refers to the cutting edge near the screw tip) impact bit bores holes from $\frac{3}{8}$ to $1\frac{1}{16}$ inches in telephone or power line poles quickly, easily, and safely, using an impact wrench. Two sizes fit all impact wrenches with the use of $\frac{7}{16}$- or $\frac{5}{8}$-inch hexagonal adapters. Overall length is 12, 16, or 24 inches, depending upon the twist length (8, 12, or 18 inches) needed. The diameter of the drill can be $\frac{3}{8}$ to $1\frac{1}{16}$ inches, depending upon the bit chosen.

Fig. 7-43 *A single-spur impact-wrench bit.* (Greenlee)

Rafting auger   This is sometimes referred to as a *boom auger*. This tool has a double twist with a flat-cut pattern head (Fig. 7-44). It is available in overall length of 14 to 20 inches, with hole diameters ranging from $1\frac{1}{4}$ to 4 inches.

Fig. 7-44 *A rafting auger.* (Greenlee)

Ship-auger bit   It has a shank for use with $\frac{3}{8}$-inch and larger electric drills. Electricians, plumbers, and other skilled workers employ it to drill deep holes or where extra reach is needed (Fig. 7-45). The ship auger has a medium-feed screw point and cutting head, with side lip to withstand rough usage. It can be used with a power bit extension to increase its length. Overall length is about 17 inches normally, and it can be had in sizes from $\frac{3}{8}$ to 1 inch.

Fig. 7-45 *A ship-auger power bit.* (Greenlee)

Extensions   A *brace-bit extension* (Fig. 7-46) is used to extend the distance from the drill bit to the power source. It allows an electrician to get at places otherwise inaccessible because it allows for operation of the brace in a restricted area.

Fig. 7-46 *A brace-bit extension.* (Greenlee)

Power-bit extensions are needed to get at some locations where the power drill and normal drill bit could not. The bit extension in Fig. 7-47 has one flat side with two setscrew holes permitting two or more extensions to be locked together for longer reach. They come in either 18- or 24-inch lengths.

Fig. 7-47 *A power-bit extension.* (Greenlee)

The *power-bit expansion* is an extension with a $\frac{1}{4}$-inch hex end to take $\frac{1}{4}$-inch flat *zip bits* (Fig. 7-48). The hex configuration makes for a secure bit fit in the extension. This one is available with an overall length of 12 inches.

Fig. 7-48 *A power-bit expansion for use with hex shank bits.* (Greenlee)

The previous list is by no means the extent of drill bits available. This has been an effort to illustrate some of the available types, and to show the difference between single-twist, double-twist, and solid-center bits. The spur side, or cutting edge, may make a difference in the job or the finish of the hole. The rate of feed and the ease with which a hole can be bored are also important with a hand-operated brace.

Extensions are basically of three types. They may vary by having either hand- or power-driven, or round or hexagonal shanks.

Special Equipment

Nail puller   The quick, easy removal of nails is important to an electrician. Wooden boxes, crates, or old construction frequently require nails removed so that the electrician can get the job done (Fig. 7-49).

Fig. 7-49 *A nail puller.* (Greenlee)

The impact handle slides on an upright claw bar for quick hammer action, driving the claw under the nailhead. The claw hook is extended to form a leverage foot and to guide the handle for fast, safe positioning of claws, and to provide maximum pull on the nail.

Note: Table 7-1 lists a few conversions that may prove helpful in converting some of the customary measurements in this chapter.

Table 7-1 *Metric Measurements**

1 inch	= 25.4 mm
1 foot	= 0.3048 m
1 sq in	= 6.4516 cm^2
1 sq ft	= 0.09290304 m^2
1 cu in	= 16.387064 cm^3
1 cu ft	= 0.028316846592 m^3

*These units are given here so that the measurements in this chapter can be converted to metric if the occasion arises.

CHAPTER

House Service
and Circuits

FOR AN ELECTRICAL SYSTEM TO OPERATE PROPERLY, it is necessary to design the system so that the proper amount of power is available. This requires the house to be wired so that wire of the proper size serves each plug in the house.

Most power is distributed locally within a neighborhood by overhead wires (Fig. 8-1). The wires are usually located in the rear of the house. In some communities the wires are buried and come into the house near the basement or foundation wall. Power is brought from the pole or transformer into the rear of the house by one black, one red, and one white (uninsulated) wire (Fig. 8-2). The wires may be three separate ones, or they may be twisted together to look like one cable (Fig. 8- 3).

Once the cable is connected to the house, it is brought down to the meter by way of a sheathed cable with three wires. These are red, white, and black. The white wire is usually uninsulated and twisted as shown in Fig. 8-4. The uninsulated wire is the ground or neutral.

Figure 8-5 shows how the power is fed from the transformer to the distribution box.

SERVICE ENTRANCE

All services should be three-wire. The capacity of service entrance conductors, and the rating of service equipment, should not be less than that shown in Table 8-1. This is a quick way to determine the type of service required. There are more accurate ways. These will be explained later in this chapter.

Fig. 8-2 *A transformer on a pole located at the rear of the house. Note the high-voltage line on top and the low-voltage (240-volt) line on the bottom, with a takeoff for the house. The bottom line is the cable television coaxial cable.*

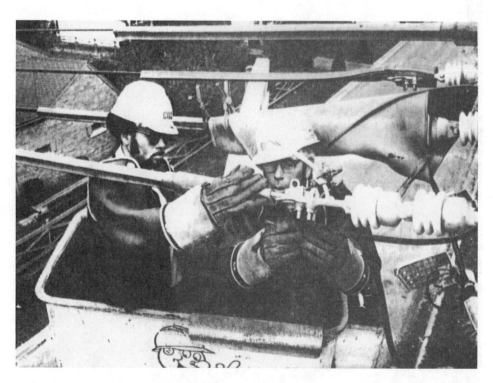

Fig. 8-1 *Stringing overhead lines for local service in a residential neighborhood.* (Cilco)

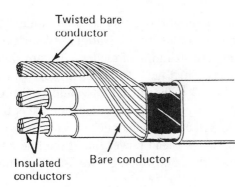

Twisted bare
conductor

Insulated
conductors

Bare conductor

Fig. 8-3 *Note how the twisted bare conductor is made from the outside coating of the other two wires.*

TYPE SE –

Fig. 8-4 *Type SE cable. This cable has three wires: one red, one black, and one uninsulated wire that forms a protective coating for the other two.*

Table 8-1 *Minimum Service Capacities for Various House Sizes*

Floor Area		Minimum Service Capacity, A
m²	ft²	
Up to 93	Up to 1000	125
93–186	1001–2000	150
186–279	2001–3000	200

These capacities are sufficient to provide power for the following:

- Lighting
- Portable appliances
- Equipment for which individual appliance circuits are required
- Electric space heating of individual rooms
- Air conditioning

A larger service may be required for larger houses. It may also be required if a central furnace or central hydronic boiler is used for electric space heating.

Many major appliances in the kitchen require individual equipment circuits. Thus, where practicable, electric service equipment should be located near or

(A)

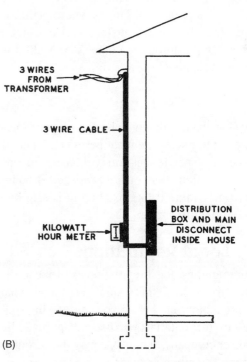

3 WIRES FROM TRANSFORMER

3 WIRE CABLE

KILOWATT HOUR METER

DISTRIBUTION BOX AND MAIN DISCONNECT INSIDE HOUSE

(B)

Fig. 8-5 *Power from the transformer is fed through the wires and down the side of the house to the meter. Finally it goes through the wall at a lower level to a distribution box in the basement.*

on a kitchen wall. This will minimize installation and wiring costs. In addition, such a location will often be convenient to the laundry. This will minimize circuit runs to laundry appliances.

Prepare Your Plans

The first step in wiring a house is planning. This means you should make accurate plans and know exactly how much material you will need. Once you have learned about the boxes, wire, and tools, then you can take a close look at the Codes.

The local power company can advise you regarding the kind of service to use for your house. In most instances, the local power company will take the power right up to the house. The rest of the wiring is your responsibility.

The electrician works on a house at two of the construction phases. The electrician is needed to put in the wiring while the house is still in the roughed-in stage. This means a new house will have the wiring installed before the drywall goes onto the studs. Once the drywall workers and carpenters have finished their work, the electrician needs to return. At this time the electrician will install the plugs, receptacle faceplates, circuit breakers, chandeliers, and lights. Switches will be connected and circuits tested for proper operation. In some cases the electrician will have to install the furnace, the electric heating, or the electric range. Various stages of construction will make it obvious that certain wiring should be done in an area before it is closed up. This is where planning is very important. It is much easier to wire a house while it is still unplastered than when it is walled in and ready for occupancy.

Permits

Before you start the installation, check with your local power company to make sure what permits are necessary. Most permits specify that you will have someone inspect the installation after you have finished. This inspection will determine if the installation is safe and free from possible fire hazards.

Local Regulations

In some communities, local regulations supersede the National Electric Code. Make sure you know what these local regulations are before you start the installation. Make sure the materials you use are approved by the local power company. If not, you may have trouble getting service once the job is done.

Some regulations apply as a minimum. Those provided as guides by the National Electrical Code relative to the stringing of feeder wires are similar to those shown in Fig. 8-6. Here, a roof with a rise greater than 4 inches in 1 foot would be difficult to walk on easily. Therefore, one of the exceptions to the Code will allow a clearance of 3 feet between the roof and an overhead power line if the line does not carry more than 300 volts. Figure 8-6 illustrates such a situation.

The electrician's main concern is the actual service drop and its entry into the house. The location of the entrance is important. The Code does have something to say about where the service can be located. Look at Fig. 8-7. Note that the service is required to be at least 3 feet from a window.

The possibility of rot caused by water can be minimized by placing the service head as shown in Fig. 8-8. The installation of this service head is usually the responsibility of the residential electrician.

Service from Head to Box

The service wires are not connected to the power company's lines until the inside of the house is properly wired and inspected. Figure 8-9 shows how the service is connected and the meter socket is inserted in the line. The overhead connections can be made if the meter is left out of the socket. This causes the rest of the service to be inoperative until the meter is properly inserted in the socket by the power company.

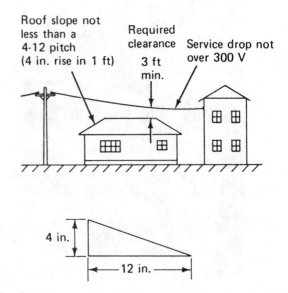

Fig. 8-6 *Clearance required for a 4-12 pitch roof in the path of a power line.*

Note how the neutral wire is also connected to the *cold* water pipe inside the house (Figs. 8-10 and 8-11). In some cases the house does not have a cold water system furnished from a local community water source. If a well is used for water, the neutral wire must be grounded as shown in Fig. 8-12.

A conduit might be used for the installation of the service into the house. Here, three wires are used. One is black, one is white, and one is red. The size of the wires will vary according to the amount of current you will be needing. This will be covered later in this chapter.

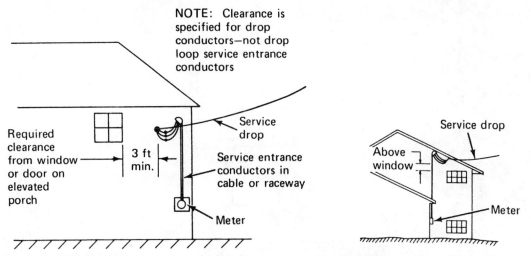

Fig. 8-7 *Minimum clearances for windows and doors for the installation of a service drop.*

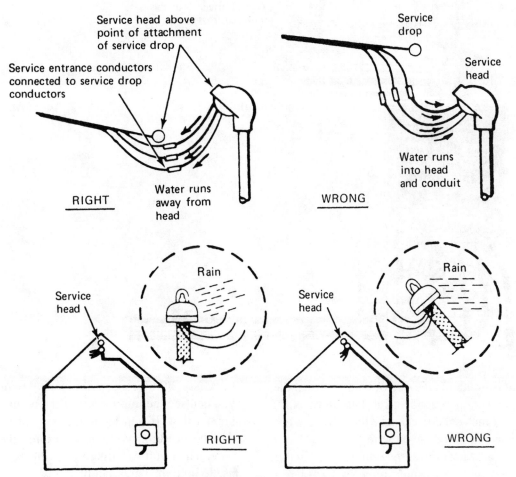

Fig. 8-8 *Proper and improper ways of installing a service head and service drop.*

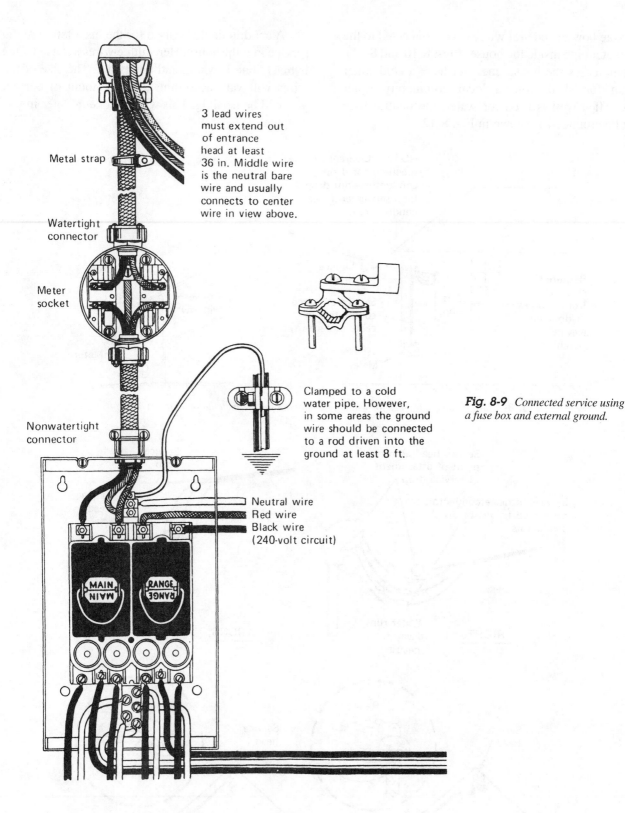

3 lead wires must extend out of entrance head at least 36 in. Middle wire is the neutral bare wire and usually connects to center wire in view above.

Metal strap

Watertight connector

Meter socket

Nonwatertight connector

Clamped to a cold water pipe. However, in some areas the ground wire should be connected to a rod driven into the ground at least 8 ft.

Neutral wire
Red wire
Black wire
(240-volt circuit)

MAIN

RANGE

Fig. 8-9 *Connected service using a fuse box and external ground.*

Conduit is limited in the size of wire it can handle. For instance, a ³/₄-inch conduit can take three No. 8 wires. The 1¹/₄-inch conduit can take three No. 2 wires, three No. 3 wires, three No. 4 wires, or three No. 6 wires. A 1¹/₂-inch conduit can take three No. 1 wires. The 2-inch conduit will handle three No. 1/0 wires, three No. 2/0 wires, or three No. 3/0 wires. Therefore, you can see the importance of the requirements being known first. If you know the amount of current you will be needing, you can determine the size of the wire. Then you can figure out which size conduit to use to hold the wires safely.

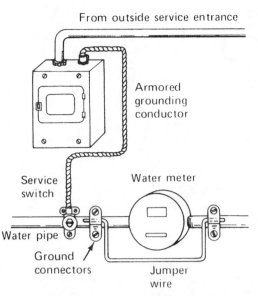

Fig. 8-10 The usual method of grounding city and town systems.

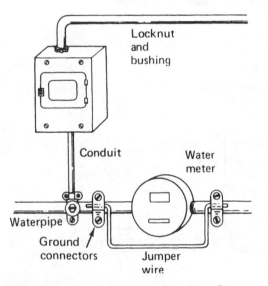

Fig. 8-11 The usual method of grounding city and town systems using conduit.

Wires from the utility company's pole to your building are called the *service drop*. These are usually furnished by the company. These wires must be high enough to provide proper clearance above grade. They also must not come within 3 feet of doors, windows, fire escapes, or any opening. See Fig. 8-7. The structure to which the service drop wires are fastened must be sturdy to withstand the pull of ice and wind.

Installation of the Service

Figure 8-13 shows how the conduit for the service is connected to the head and to the meter socket. Note how it enters the house with a waterproof fitting and then goes unspliced into the distribuiton panel in the basement.

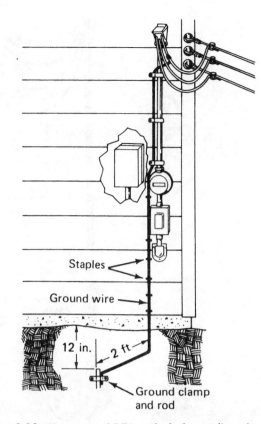

Fig. 8-12 The approved REA method of grounding wire system with ground rod.

Circuits are run from the distribution panel to various parts of the house. Some examples of circuits are shown in Fig. 8-13. Note that here there are 22 circuits available. Figure 8-9 shows how the fuse-type box is served by the entrance wires.

Conduit is fastened to the wall of the building by clamps placed every 4 feet. Note that the conduit should use an entrance ell to turn the conduit into the house. Such an ell has two threaded openings corresponding to conduit size. Use an adapter to fasten the conduit into the threaded opening at the top of the ell. Into the lower opening, fasten a piece of conduit to run through the side of the house. Use a connector to attach the conduit to the electric service panel.

Inserting Wire into a Conduit

After the conduit is installed, you can push the wires through the top hub of the meter, through the conduit, and out of the service head. All three wires must extend at least 36 inches out of the service head. This will allow enough wire for connecting to the power lines. Then the wires are brought down from the meter to the entrance ell. Remove the ell cover and pull the wires through to the service panel inside the house. Use a white-covered insulated wire for the neutral in the conduit.

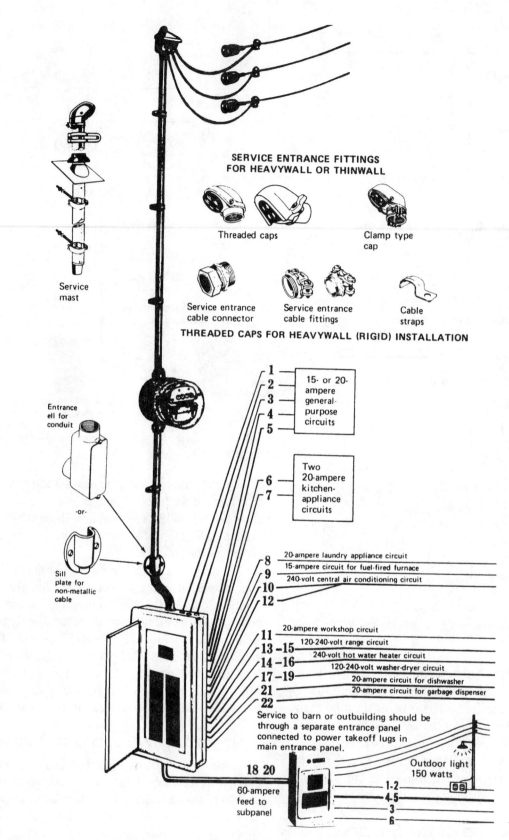

SERVICE ENTRANCE FITTINGS
FOR HEAVYWALL OR THINWALL

Threaded caps

Clamp type
cap

Service entrance
cable connector

Service entrance
cable fittings

Cable
straps

THREADED CAPS FOR HEAVYWALL (RIGID) INSTALLATION

Service
mast

Entrance
ell for
conduit

-or-

Sill
plate for
non-metallic
cable

1
2
3
4
5

15- or 20-
ampere
general-
purpose
circuits

6
7

Two
20-ampere
kitchen-
appliance
circuits

8 20-ampere laundry appliance circuit
9 15-ampere circuit for fuel-fired furnace
10 240-volt central air conditioning circuit
12

11 20-ampere workshop circuit
13 –15 120-240-volt range circuit
14 –16 240-volt hot water heater circuit
17 –19 120-240-volt washer-dryer circuit
21 20-ampere circuit for dishwasher
22 20-ampere circuit for garbage dispenser

Service to barn or outbuilding should be
through a separate entrance panel
connected to power takeoff lugs in
main entrance panel.

Outdoor light
150 watts

18 20

60-ampere
feed to
subpanel

1-2
4-5
3
6

Fig. 8-13 Service entrance.

DISTRIBUTION PANELS

There are many types of distribution panels made for the home. One type has fuses that screw in. These fuses must be replaced when they are blown, or open (Fig. 8-14). Another type is the circuit breaker box. A circuit breaker can be reset if the device is tripped by an overload. This type has gained in popularity for home use. Many people do not want to be bothered with looking for a new fuse every time one blows. The circuit breaker may be reset by pushing it to the off position and then to the on setting. If it trips a second time, the circuit trouble is still present. It should be located and removed before resetting the breaker arm again (Fig. 8-15).

For a closer look at the fuse replacement or circuit breaker, take a look at Fig. 8-16. This is a 100-ampere fuseless service entrance panel. (*Note:* This size is almost obsolete today.)

ROMEX CABLE

Romex cable is used to carry power from a distribution panel box to the individual outlets within the house. This nonmetallic sheathed cable has plastic insulation covering the wires to insulate it from the environment. Some types of cable may be buried underground. Fig-

ures 8-17 and 8-18 show the types most commonly used in house wiring.

Most new homes are wired with 12/2, or No. 12 wire with two conductors. One of the wires is white and the other is black. Inside the cable is an uninsulated ground wire of the same size, No. 12. In the past, the smallest wire size used in homes was No. 14, with 2 conductors (written on the cable as 14/2 with ground or 14/2WG). Single conductors may be used in conduits as stranded or solid conductors (Fig. 8-19). Figure 8-20 shows how the wire is cut to remove the insulation without scoring the metal part of the wire.

Fig. 8-15 *A distribution box with circuit breakers.*

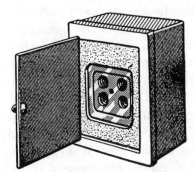

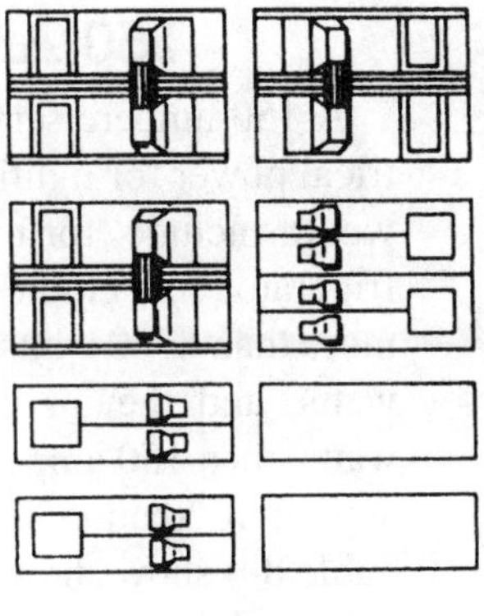

Fig. 8-14 *A distribution box. This is a four-circuit box with screw-in fuses. It is used for an add-on in some older systems.*

100-ampere main breaker (shuts off all power)

30-ampere circuit (240-volt) for dryer, hot-water heater, central air conditioning, etc.

Four 20-ampere circuits for kitchen and small appliances and power tools.

40-ampere circuit (120- to 240-volt) for electric range.

Four 15-ampere circuits for general-purpose lighting, television, vacuum cleaner.

Space for four 120-volt circuits to be added for future loads as needed.

Fig. 8-16 *An example of a fuseless service entrance panel capable of handling a 100-ampere service.*

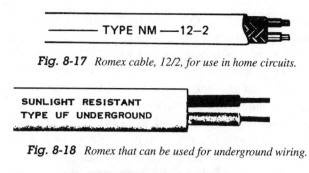

Fig. 8-17 *Romex cable, 12/2, for use in home circuits.*

Fig. 8-18 *Romex that can be used for underground wiring.*

Fig. 8-19 *Single wire, stranded, with plastic coating. This can be used in conduits. It also comes in a solid copper wire.*

Fig. 8-20 *The right way and wrong way to cut insulation from a piece of wire.*

WIRE SIZE

The larger the physical size of the wire, the smaller the number. For example, No. 14 is smaller than No. 12. Number 14 has been used for years to handle the 15-ampere home circuit. Today, the Code calls for No. 12 for a 15-ampere circuit with aluminum or copper-clad aluminum wire. This is a safety factor. Aluminum wire cannot handle as much current as copper.

Figure 8-21 shows how the size and numbers relate in determining wire diameter. Table 8-2 lists the

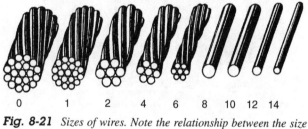

Fig. 8-21 *Sizes of wires. Note the relationship between the size and number.*

Table 8-2 Conductor Capacities for Household Equipment

Item	Conductor Capacity
Range (up to 21-kW rating) *or*	50 A–3 W–115/230 V
{Built-in oven	30 A–3 W–115/230 V
{Built-in surface units	30 A–3 W–115/230 V
Combination washer-dryer *or*	40 A–3 W–115/230 V
electric clothes dryer	30 A–3 W–115/230 V
Fossil-fuel-fired heating equipment (if installed)	15 A or 20 A–2 W–115 V
Dishwasher and waste disposer (if necessary plumbing is installed)	20 A–3 W–115/230 V
Water heater (if installed)	Consult local utility

capacities of wires needed for various household equipment. Table 8-3 shows conductor capacities for other household equipment. You should consider this information whenever you are planning an installation.

Table 8-3 Conductor Capacities for Other Household Equipment

Item	Conductor Capacity
Room air conditioners *or*	20 A–2 W–230 V
central air conditioning unit *or*	40 A or 50 A–2 W–230 V
attic fan	20 A–2 W–115 V (Switched)
Food freezer	20 A–2 W–115/230 V
Water pump (where used)	20 A–2 W–115/230 V
Bathroom heater	20 A–2 W–115/230 V
Workshop or workbench	20 A–3 W–115/230 V

PLANNING THE RIGHT SIZE SERVICE AND CIRCUITS

Most houses have either a 150-ampere or a 200-ampere service from the pole in the back of the house to the circuit breaker box in the basement. Number 1/0 or No. 3/0 (type RHW insulation) three-wire is usually used. This usually provides sufficient power for household needs. One way to find out how much power is needed is to list the devices that use power in a house and how much. Add up the requirements. Divide the wattage by 120 to find out how many amperes would be needed at any time.

In a house with an electric range, water heater, high-speed dryer or central air conditioning, together with lighting and the usual small appliances, there is at least a 150-ampere requirement. If the house also has electric heating, a 200-ampere line should be installed.

150-Ampere Service

A 150-ampere service will provide sufficient electrical power for lighting and portable appliances. This would include, for example, a roaster, rotisserie, refrigerator, and clothes iron. If the dryer does not draw more than 8700 watts, the range no more than 12 kilowatts, and the air conditioner no more than 5000 watts, then 150 amperes will be sufficient. Do not add any more than 550 watts of appliances, however. Table 8-4 shows how much power is required to operate small devices.

Table 8-4 *Appliances and Their Wattages*

Device	Wattage	Voltage	Current
Small Appliances			
Blender	950	120	7.92
Clothes iron	1,100	120	9.17
Electric fryer (small)	1,200	120	10.00
Food processor	400	120	3.34
French fryer	900	120	7.50
Hair dryer	1,200	120	10.00
Hand drill (1/3-hp)	324	120	2.70
Microwave oven	1,680	120	14.00
Mixer	165	120	1.38
Toaster	1,200	120	10.00
Vacuum cleaner (1-hp)	746	120	6.22
Waffle grill	1,400	120	11.67
Large Appliances			
Clothes dryer	5,000	240	20.83
Dishwasher	1,200	120	10.00
Laundry circuit	3,000	240	12.50
Range	12,000	240	50.00
Space heating	9,000	240	37.50
Water heater	2,500	240	10.40

Branch Circuits

General-purpose circuits should supply all lighting and all convenience outlets, except those in the kitchen, dining room (or dining areas of other rooms), and laundry rooms. General-purpose circuits should be provided on the basis of one 20-ampere circuit for not more than every 500 square feet or one 15-ampere circuit for not more than every 375 square feet of floor area. Outlets supplied by these circuits should be divided equally among the circuits (Figs. 8-22 through 8-30).

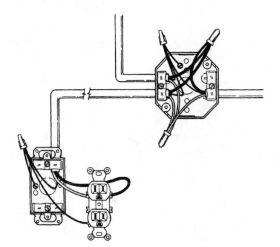

Fig. 8-23 *To add a convenience outlet from an existing junction box.*

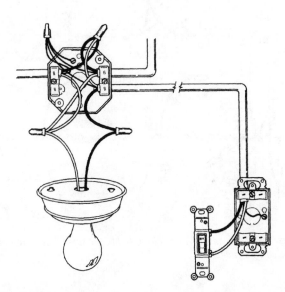

Fig. 8-24 *To add a wall switch to control a ceiling light in the middle of a run.*

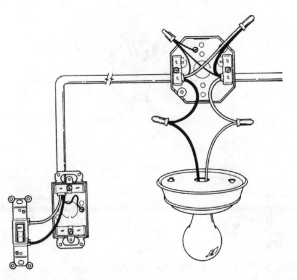

Fig. 8-22 *To add a wall switch for control of a ceiling light at the end of the run.*

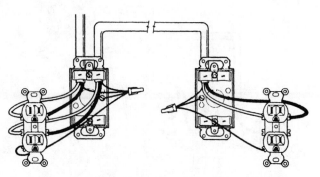

Fig. 8-25 *To add new convenience outlets beyond old outlets.*

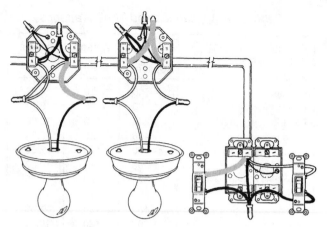

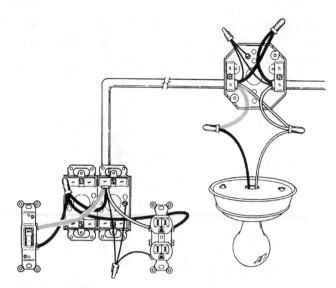

Fig. 8-26 *To add a switch and convenience outlet in one box, beyond existing ceiling light.*

Fig. 8-29 *To install one ceiling outlet and two new switch outlets from existing ceiling outlet.*

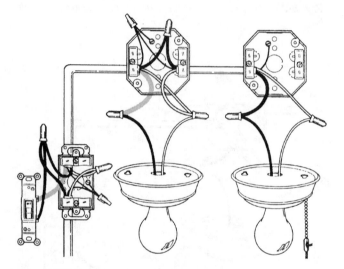

Fig. 8-27 *To install two ceiling lights on same line, one controlled by switch.*

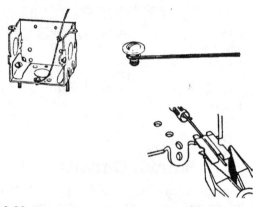

Fig. 8-30 *Two methods of attaching ground wire to metal box.*

These requirements for general-purpose branch circuits take into consideration the provision in the current edition of the National Electrical Code. Floor area designations are in keeping with present-day use of such circuits (Figs. 8-31 through 8-39), (pages 122-124).

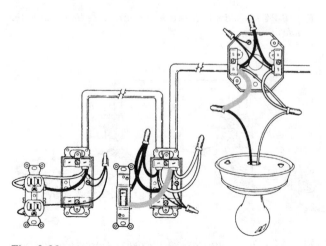

Fig. 8-28 *To add a switch and convenience outlet beyond existing ceiling light.*

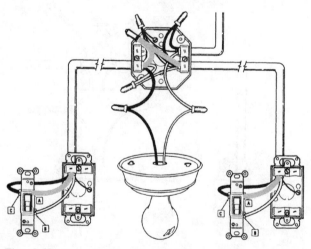

Fig. 8-31 *Here, the three-way switches control a single lamp. Feed is through the center-lamp box. The terminals marked A and B are the light-colored points to which red and white wires must be connected. Terminal C is the dark-colored (brass screw) point to which the black wire must be connected.*

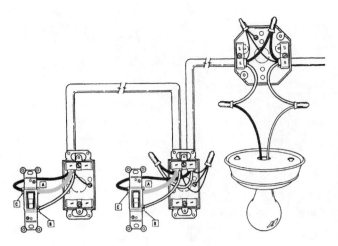

Fig. 8-32 *Two three-way switches are tied together with a three-wire Romex cable. The lamp and switch box are fed with a two-wire cable. The terminals marked A and B are the light-colored screws to which red and white wires must be connected. Terminal C is the dark-colored (brass-colored screw) point to which the black wire must be connected.*

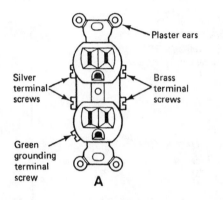

SINGLE - POLE

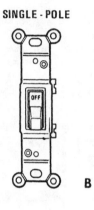

Fig. 8-33 *Switching arrangements for (B) single-pole, (C) three-way, and (D) four-way switches. Note the switches and their terminal screws. Also note the terminals on the receptacle (A).*

THREE - WAY

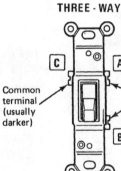

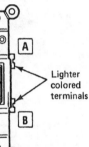

FOUR - WAY

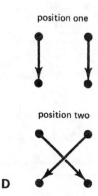

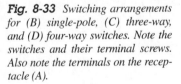

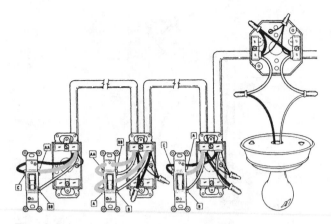

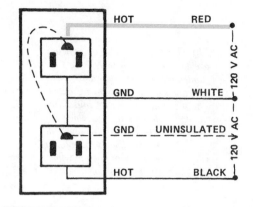

Fig. 8-36 *Three wires feeding a single receptacle.*

Fig. 8-34 *Two three-way switches and one four-way switch make it possible to control a lamp from three locations. Terminals marked A and B are the light-colored screws to which red and white wires must be connected. Terminal C is the dark- colored screw to which the black wire must be connected. Terminals AA show where the two ends of the red wire are connected between the four-way and the three-way switches on the right. Terminals BB show where the two ends of the white wire are connected to the four-way and three-way switches.*

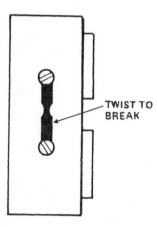

Fig. 8-37 *A breakaway connection between the two outlets allows the red and black hot wires to be connected to the same duplex outlet.*

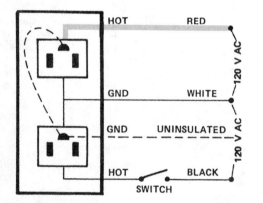

Fig. 8-35 *Three wires feeding a single receptacle. Switched lower outlet—top outlet hot all the time.*

For a close look at the internal wiring of a house, before it is covered with drywall or plaster, see Figs. 8-40 through 8-45 (pages 125–126). Wiring of switches and receptacles and the locations within the

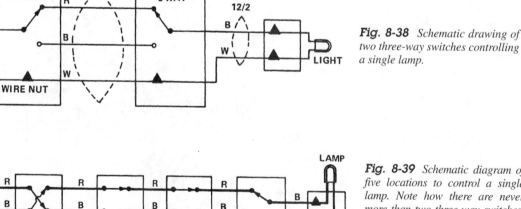

Fig. 8-38 *Schematic drawing of two three-way switches controlling a single lamp.*

Fig. 8-39 *Schematic diagram of five locations to control a single lamp. Note how there are never more than two three-way switches. Each time another location for control is added, a four-way switch is used.*

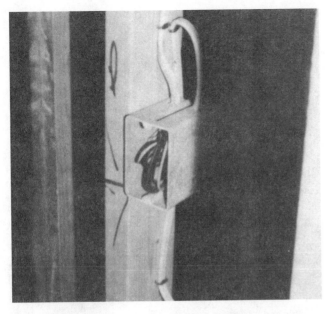

Fig. 8-40 *The cable in this switch box is fed from the top. It is then brought out from the top and stapled. Next, it leads down and is stapled again below the box. Since this is an uninsulated box, it has no clamp or connector. The staple above the box serves as a strain relief.*

circuits are shown in Figs. 8-46 through 8-55 (pages 127–129).

It is recommended that separate branch circuits be provided for both lighting and convenience outlets in living rooms and bedrooms. It is also recommended that the branch circuits servicing convenience outlets in these rooms be of the three-wire type, equipped with split-wired receptacles.

Space Heating and Air Conditioning Circuits

Electric space heating The capacity required for electric space heating should be determined from Table 8-5. The table shows maximum winter heat loss, based on the total square feet of living space in the home.

If electric space heating is installed initially, wiring should be as follows:

- For a central furnace, boiler, or heat pump: a three-wire, 115/230-volt feeder, sized to the installation.

(A)

(B)

Fig. 8-41 *(A) Location of an outlet where there is no insulation. This is the basement entrance, and the switch box is located on the top right of the photo. It will serve to turn the basement light on and off, for light on the steps. Note how the cables have been stapled about 6 to 8 inches above the box and about 36 inches above that. The cable is held approximately in the middle of the stud. (B) Three-ganged switchbox with Romex.*

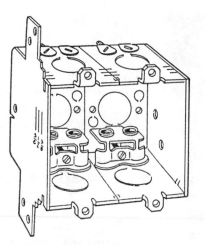

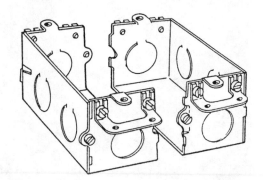

Fig. 8-42 *Three-ganged switch boxes. Note how the wires are stapled to anchor them and how each box has a cable entering and leaving. This will be a three-switch control center. See (B) of Fig. 8-41 on page 125.*

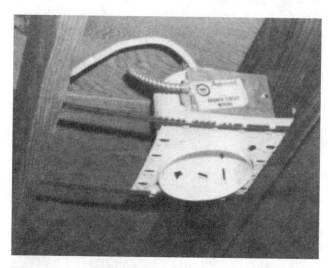

Fig. 8-43 *Note that the Romex feeds the box. BX (Cable sheathed in metal armor) is used to protect the asbestos-covered wires from the box to the lamp socket.*

Fig. 8-45 *Wires left through the siding of a new house, to be attached to outlet boxes later. The boxes are covered at present. They will have the siding removed and connections made to the lamp fixtures, and then covered and secured, making it a waterproof box.*

Fig. 8-44 *Note that the studs are drilled and the wires are fed to the switch boxes.*

- For individual room units: ceiling cable, or panels, sufficient 15-, 20-, or 30-ampere, two wire, 230-volt circuits to supply the heating units in groups, or individually (Figs. 8-56, 8-57, and 8-58).

Electric air conditioning   The capacity required for electric air conditioning is to be determined from the chart of maximum allowable summer heat gain, based on the total square feet of living space in the home. See Table 8-6.

If electric air conditioning is installed initially, wiring should be provided for the following:

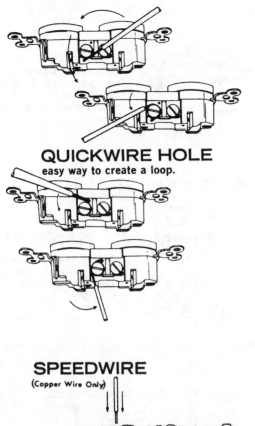

QUICKWIRE HOLE
easy way to create a loop.

SPEEDWIRE
(Copper Wire Only)

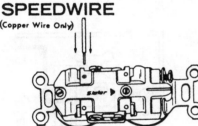

Fig. 8-46 *Duplex receptacles being wired. A speed-wire connection means that the wire is pushed into the hole to make contact.*

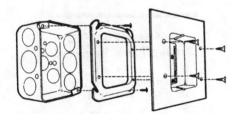

Fig. 8-47 *Recessed outlet mounted in a 4-inch-square box.*

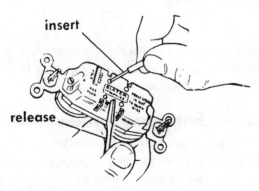

insert

release

Fig. 8-48 *Note how the wire is released by pressing with a screwdriver blade into the slot.*

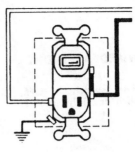

COMMON FEED ONLY

When plug is inserted into receptacle, pilot light automatically lights up.

Fig. 8-49 *Receptacle with pilot light.*

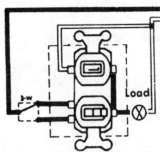

WIRED AS PILOT LIGHT

When 3-Way switch is ON, pilot light glows. This circuit requires a ground wire for this operation.

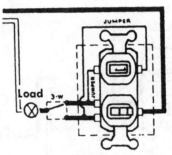

WIRED AS NIGHT LIGHT (Breakoff Tab Removed)

When 3-Way switch is OFF, indicator light glows.

Fig. 8-50 *Switch and pilot light.*

- A heat pump, providing both winter heat and summer air conditioning; a three-wire, 115/230-volt feeder, sized to the installation.

- Individual room air conditioning units, sufficient 15-, 20-, or 30-ampere, three-wire, 230-volt circuits to supply all units and a 20-ampere, 230-volt, three-wire outlet in each room, on an outside wall and convenient to a window.

In some cases, neither electric space heating nor electric air conditioning equipment is installed initially. In such an instance, the service entrance conductors and service equipment should have the capacity required, in accordance with the appropriate chart. Spare feeder or circuit equipment should be provided, or provision for these should be made in panelboard bus space and capacity.

The plan should allow bus space and capacity for a feeder position that can serve a central electric heating

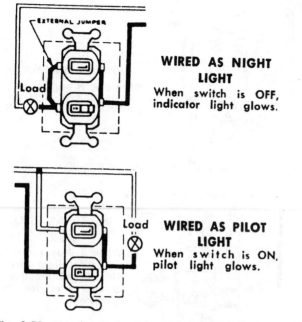

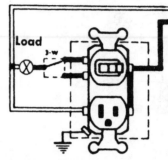

WIRED AS NIGHT LIGHT

When switch is OFF, indicator light glows.

WIRED AS PILOT LIGHT

When switch is ON, pilot light glows.

Grounded power outlet controlled from 3-Way switch and 3-Way switch at another location.

COMMON FEED

3-Way switch controls light only. Power outlet for grounded appliances.

Fig. 8-51 *Switch and pilot light device wired with a three-way switch.*

Fig. 8-52 *Power outlet and switch can be used in a number of configurations for circuit control.*

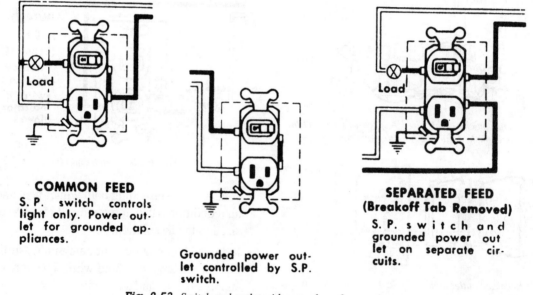

COMMON FEED

S.P. switch controls light only. Power outlet for grounded appliances.

Grounded power outlet controlled by S.P. switch.

SEPARATED FEED
(Breakoff Tab Removed)

S.P. switch and grounded power outlet on separate circuits.

Fig. 8-53 *Switch and outlet with a number of arrangements.*

or air conditioning plant directly, or can supply a separate panelboard to be installed later. The panelboard would control circuits to individual room heating or air conditioning units. Space heating and air conditioning are considered dissimilar and noncoincident loads. Service and feeder capacity need be provided only for the larger load, not for both.

Space heating and air conditioning outlets Since many different systems and types of equipment are available for both space heating and air conditioning, it is impractical to show outlet requirements for these

uses in each room. Electrical plans and specifications for the house should indicate the type of system to be supplied and the location of each outlet.

Entrance Signals

Entrance push buttons should be installed at each commonly used entrance door and connected to the door chime. They should give a distinctive signal for both front and rear entrances. Electrical supply for entrance signals should be obtained from an adequate bell-ringing or chime transformer.

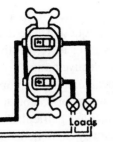

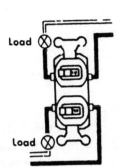

COMMON FEED

Two S.P. switches on same circuit. Each switch controls an independent light.

SEPARATED FEED
(Breakoff Tab Removed)

Two S.P. switches on separate circuit. Each switch controls an independent light.

Fig. 8-54 Two switches mounted in one unit with possibilities for controlling two loads in different ways.

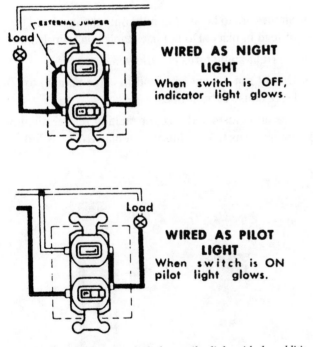

WIRED AS NIGHT LIGHT

When switch is OFF, indicator light glows.

WIRED AS PILOT LIGHT

When switch is ON pilot light glows.

Fig. 8-55 Switch and night-light or pilot light with the addition of an external jumper.

Table 8-5 *Maximum Heat Loss Values**
(Based on Infiltration Rate of ¾ Air Change/Hr)

Degree-days	Watts		Btu · h	
	m²	ft²	m²	ft²
Over 8000	0.98	10.6	3.3	36
7001 to 8000	0.93	10.0	3.2	34
6001 to 7000	0.88	9.5	3.0	32
4501 to 6000	0.85	9.2	2.9	31
3001 to 4500	0.83	8.9	2.8	30
3000 and under	0.78	8.4	2.7	29

*For new homes. May be exceeded in converting existing homes.

Fig. 8-56 Baseboard electric heater.

Fig. 8-57 Electric heater showing thermostat.

A voice intercommunications system permits the resident and a caller to converse without opening the door. This is convenient and is an added protection. It may be designed for this purpose alone, or it may be part of an overall intercommunications system for an entire house.

In a smaller home, the door chime is often installed in the kitchen, providing it will be heard throughout the home. If not, the chime should be installed at a more central location, usually the entrance hall. In a larger home, a second chime is often needed. This will ensure that it is heard throughout the living quarters.

Entrance-signal conductors should be no smaller than the equivalent of a No. 18 AWG copper wire. (AWG is the abbreviation for American Wire Gage.)

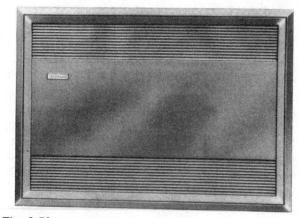

Fig. 8-58 *Smooth-finish, flush installation of an electric heater.*

Table 8-6 *Maximum Allowable Heat Gain (Btu · h)**

Area to Be Conditioned		Design, Dry-Bulb Temperature†			
m²	ft²	32°C (90°F)	35°C (95°F)	38°C (100°F)	40°C (105°F)
70	750	15,750	18,000	19,500	21,000
93	1,000	20,500	23,000	24,750	26,500
139	1,500	27,000	30,000	31,500	33,000
186	2,000	36,000	40,000	42,000	44,000
232	2,500	45,000	50,000	52,500	55,000
279	3,000	54,000	60,000	63,000	66,000

*Based on FHA minimum property standards.

†Based on dry-bulb temperature of less than 32°C (90°F); use 32°C (90°F) values. For designed dry-bulb temperature exceeding 40°C (105°F), use 40°C (105°F) values. (To find °C, simply subtract 32 from °F and multiply by 5/9.)

Communications

An intercommunications system, with a station in each room of the living quarters, is recommended. It may be arranged for voice communication to the entire house, or selectively to an individual room only. Music from AM or FM radio, stereo, or recordings may be fed into the system. It may also be used to communicate with a caller without opening the door.

For the larger home with accommodations for a resident servant, a dining room–to–kitchen signal is also recommended. This can be operated by a push button attached to the underside of the dining table or by a floor tread placed under the table. In some cases a built-in annunciator in the kitchen, with push-button stations in various rooms of the living quarters, or a separate intercommunication telephone system is desirable.

Intercommunication systems and telephones should be operated from a power unit recommended by the manufacturer of the system. They should be supplied with a minimum of 115 volts.

For telephones, it is recommended that raceways and outlets be installed at the time of construction. Outlets, if desired, should be located in the kitchen, living room, bedrooms, and other rooms as requested. The telephone company installs concealed wiring in residential units during the construction, on a selective basis.

It is suggested that the local telephone company be consulted for details of service prior to construction. The company should be consulted particularly in regard to the installation of raceways and the location of protector equipment.

Television

An outdoor television antenna and lead-in connections require a nonmetallic outlet box at a convenient location. This is needed for connection of the master antenna system to the various other locations in the house. Allow for a 115-volt outlet near the antenna rotor control. It is more convenient if the antenna lead-in wire and the control circuits for the rotor are installed before the house has been closed in by drywall or by plaster. The appearance is also improved. Follow the rules and regulations for installing lead-in wire:

- Staples are to be used to hold only the insulated type of lead in place. Do not use staples on twin-lead.
- No right-angle turns are allowed.
- Do not run the antenna lead-in cable parallel to 60-hertz house current lines.
- Use an uninsulated box for making the termination connections to the plug-in receptacle (Fig. 8-59).

Fig. 8-59 *TV outlets wired into boxes with receptacle for 120 volts to power the television rotor that is mounted on the roof of the house.*

Cable TV installation is now widespread. Installation of coaxial cable to all rooms where TV will be viewed is necessary. This should be done at the same time as installation of branch circuits. The coaxial cable should be stapled to the studs and rafters as needed. Terminate the cable in a box so that a connector can be attached. Refer to NEC Article 820.50.

Radio

In their modern form, radio receivers used purely for the reception of broadcast communications seldom require

antenna and ground connections. If FM reception is desired, it is recommended that provisions similar to those for television be included.

Household Warning Systems

An automatic fire warning system is recommended, including installation of both smoke and heat detectors. A smoke detector should be located in the immediate vicinity of, but outside, the bedrooms. Additional smoke detectors placed in strategic locations around the home, and in each bedroom, are recommended.

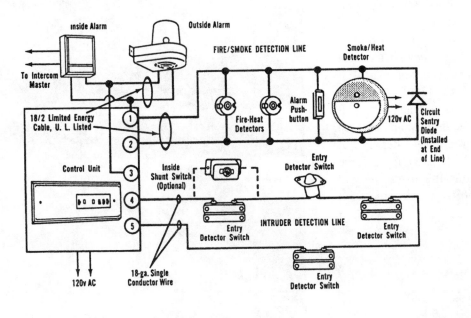

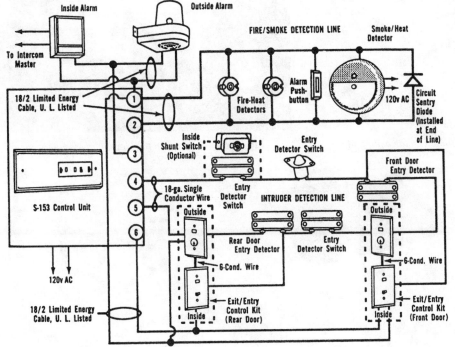

Fig. 8-60 Wiring diagrams for home burglar alarm, smoke detector, and associated devices.

Heat detectors should be provided in each room and each hall of the living quarters. They should also be provided in the attic, furnace room, utility room, basement, attached garage, and each closet and partitioned-off storage area (Fig. 8-60).

The fire alarm bell should be clearly audible in all bedrooms, even with doors closed. An outside alarm bell is also recommended. A test button should be installed in the master bedroom or other desirable location.

The primary power supply should be 115 volts, feeding a separate and independent transformer recommended by the manufacturer of the system. A "power on" visible indicator should be provided. A battery-operated standby source of power is also desirable.

Figure 8-61 pictures the newest type of fire-heat detector. It is mounted on the ceiling. It can detect heat and set off an alarm before a fire gets out of control. If the temperature rises above 57°C (135°F), the fire-heat detector automatically activates an alarm. It should be used in living areas, bedrooms, and closets. It is only 1³/₄ inches in diameter and has a depth of ³/₄ inch. Another model, which looks exactly like the 57°C (135°F) model, is available. It activates only at 93°C (200°F). This is designed for use in attics and furnace rooms primarily.

A number of different ways are needed to trigger an alarm in case of an illegal entry. These require different switches. Some are used to detect the opening of a window. Others are connected to the door. And there is a mat

Fig. 8-63 *An outside alarm bell.*

Fig. 8-64 *A siren and flashing-light warning device.*

Fig. 8-61 *A fire-heat detector mounted on the ceiling tile.*

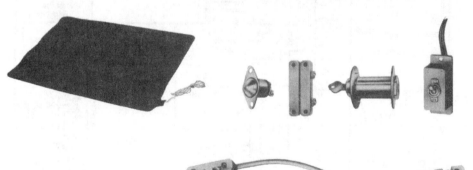

Fig. 8-62 *Burglar alarm switches.*

with a switch to detect entry through a door. Figure 8-62 shows some of the different types of switches available to work in conjunction with a burglar alarm system.

An outside bell is needed to alert neighbors (Fig. 8-63). They can call the police in case you are not at home. This bell can also be wired to the fire-smoke detector system and will awaken you in case of fire or smoke.

Sometimes a light and a siren are necessary. In this case, a model like the one in Fig. 8-64 will fill the order. It has a flashing red light as well as an audible warning.

A number of smoke detector systems are now on the market (Fig. 8-65). Most cities are requiring all buildings to have smoke detector systems. This will mean much work for electricians. When properly working, the systems can save many lives and millions of dollars each year. Many new systems will be coming on the market. It will be to your benefit to keep close check on the new methods used to detect smoke.

The systems are basically simple. Try to obtain a wiring diagram of each system so that you can repair one if necessary. Most of the present-day smoke alarm detector units use a pilot light that must be on at all times. There is a manufacturer who uses *nuclear* materials in the smoke detector unit. Get the proper information before you try to check a unit. It may be dangerous to your health. Follow instructions properly and to the letter.

Installing Doorbells and Chimes

The installation of doorbells and chimes is an easy job. The wire used is No. 18 wire. The connections are simple. Usually the wire comes in two- or three-conductor cable. This is taken from the transformer to the switch and the chime, as shown in Fig. 8-66. The other side of the transformer goes to the 120-volt line. Usually the transformer is mounted on the side of the circuit breaker box. The primary wires are fed through a hole in the box. The transformer has a screw that makes it easily secured to the box. See Fig. 8-67.

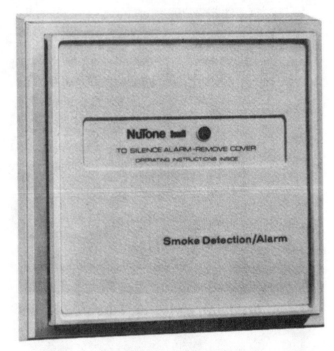

Fig. 8-65 *A smoke detector.*

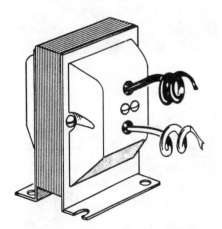

Fig. 8-67 *A doorbell or chime transformer. Note the screw near the curled wires. This screw and the ears sticking up to hold the screw fit into a knockout in the entrance panel. They tighten to hold the transformer securely against the outside of the panel.*

Fig. 8-66 *Installation of doorbells and chimes.*

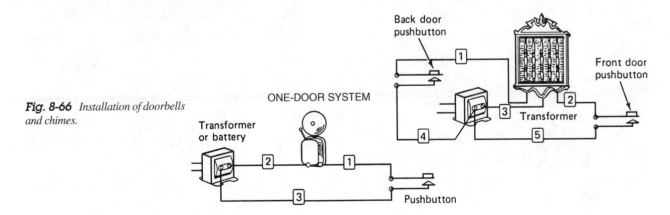

KILOWATTHOUR METERS

Meters measure power used by a household, business, or industry. There are several types of kilowatthour meters (Fig. 8-68 and 8-69). *Kilo* means 1000. *Kilowatthour* means 1000 watts of energy used for 1 hour. The unit of time for this type of meter is the hour. This means that whenever the meter reader records the meter reading—it may be a month from one reading to the other—the entire period in which power has been used is equated to 1 hour. The energy consumed is based on 1 hour. The number of watts would, in most cases, be thousands. Thus, the kilowatthour is used for measuring the power consumed, no matter what the length of time.

Sometimes the meter reader makes the rounds every 28 days, and sometimes every 31 days (Fig. 8-70). This means the reading is higher or lower in terms of usage. However, the time period for basing the consumption is still 1 hour. In a regular wattmeter, where there is no accumulative factor included, the reading and the watts consumed are in terms of the second, or what is being read at the moment the meter is used. This type of meter reading is usually done in the laboratory, or under test conditions where a definite consumption rate is necessary for a device.

In Fig. 8-71, the meter movement is composed of two coils. One coil is across the line to check the voltage. The other coil is in series with the load to check

Fig. 8-68 *A kilowatthour meter.*

Fig. 8-69 *A three-phase kilowatthour meter. This type is designed for industrial use.*

Fig. 8-70 *A meter reader for a utility company. Now, a signal from the meter along the power lines tells the power company what the present meter reading registers.*

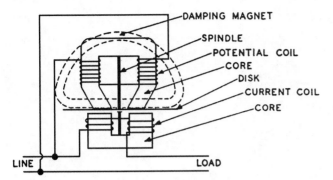

Fig. 8-71 *The parts of a kilowatthour meter and its hookup.*

the current being drawn. This arrangement accounts for the voltage times the current, converting to watts, or power consumed. Power is equal to the product of the voltage and current, or $P = E \times I$.

Reading a Kilowatthour Meter

Take a look at Fig. 8-72 and read the amount of power consumed on the meter. If the dial pointer has just passed a number, read that number and not the next-higher one. For example, in Fig. 8-72, the reading is 7432 kilowatts. Read from left to right and write down the numbers.

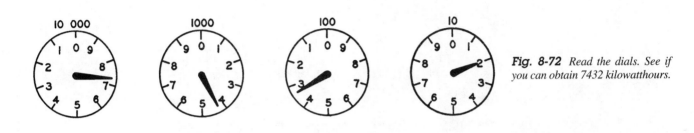

Fig. 8-72 *Read the dials. See if you can obtain 7432 kilowatthours.*

CHAPTER

Installing Romex

ROMEX

Romex cable of the No. 14 or No. 12 size is usually employed in the wiring of homes, light industries, and businesses. To utilize safely the nonmetallic sheathed cable in a manner consistent with accepted wiring practices, it is necessary to use the switch and outlet box for terminations and connections. Splices are not allowed along the run of the cable unless the splice is housed in an appropriate box and totally enclosed with a cover plate.

When insulating material is used to make boxes and other equipment for wiring a house, it is not necessary to use a clamp or connector to hold the wire in place. Such a box is permitted when the wire or cable is supported within 8 inches of the box. Cable enters the box by way of a knockout. All knockouts should be closed if not used for cable.

Nonmetallic Sheathed Cable or Romex

Nonmetallic sheathed cable Romex is one of the most widely used cables for branch circuits and feeders in residential and commercial systems. Such cable is commonly and generally called Romex by electrical construction people, even though the word Romex is a registered trade name of General Cable Corp. Industry usage has made the trade name a generic title so that nonmetallic sheathed cable made by any manufacturer might be called Romex.

This generic use of a trade name also applies to the term *BX,* which is commonly used to describe any standard armored cable made by any manufacturer—even though the term *BX* is a registered trade name of General Electric Co. Type NM cable has an overall covering of fibrous or plastic material that is flame-retardant and moisture-resistant. Type NMC is similar, but the overall covering is also fungus-resistant and corrosion-resistant. The letter C indicates that it is corrosion-resistant.

Permitted locations This type of wiring may be used either for exposed or for concealed wiring in any kind of building or structure. NM cable may be used in

- One-family dwellings
- Two-family dwellings
- Multifamily dwellings
- Other structures

But, in any case, the dwelling, building, or structure *must not have over three floors above grade.*

Type NM or NMC cable must be identified for use in cable trays. This requirement essentially calls for UL listing and marking on the cable to make it recognized as suitable for installation in trays.

Although NM cable is limited to use in "normal dry locations," NMC—the corrosion-resistant type—is permitted in "dry, damp, moist, or corrosive locations." Because it has been widely used in barns and other animals' quarters where the atmosphere is damp and corrosive (due to animal vapors), NM cable is sometimes referred to as *barn wiring.*

Box Volume

The volume of the box is the determining element in the number of conductors that can be allowed in the box. For instance, a No. 14 wire needs 2 cubic inches for each conductor. Therefore, if a two-wire Romex cable of No. 14 conductors is specified, it means that a ground wire (uninsulated) is also included which will count as a conductor, and the three wires, or conductors, will require 6 cubic inches of space. A box used for this installation should have at least this amount. A 3- × 2- × 1½-inch device box has only 9.0 cubic inches and will be allowed to contain three of No. 14, No. 12, or No. 10 conductors, but only two No. 8 conductors. Boxes manufactured recently will have the cubic inch capacity stamped on them. You should check the National Electrical Code handbook for the number of wires allowed until you are familiar with the standards for boxes.

In Figs. 9-1 through 9-14 you will find a representative sampling of devices made of insulating material

Fig. 9-1 *An insulated metallic grounding bushing.*

Fig. 9-2 *A male insulating bushing.*

Fig. 9-3 *A nonmetallic switch box for a house.*

Fig. 9-4 *A nonmetallic switch box. It will hold nine conductors of No. 1 wire, eight conductors of No. 12 wire, or seven conductors of No. 10 wire.*

that are acceptable in house wiring. Read the captions under each figure and identify the characteristics for future use in wiring buildings.

BX

BX cable, flexible metal-covered cable, is covered in detail in Chapter 10. Here, it is discussed briefly as it compares to the use of Romex in circuits for homes.

Connectors BX requires connectors that make a clamping action that holds the protective coating rigidly in place. They also make a good electrical connection for grounding purposes. The design of these connectors will vary with different manufacturers. However, the group of connectors shown in Figs. 9-15 to 9-20 will represent those made by a particular manufacturer. They will serve here as examples of connectors available for switch boxes, utility boxes, and both square and octagonal outlet boxes.

A *bushing* is needed to protect the wires inside the armored cable from abrasion and from cuts in the insulation. Note that the connector is mounted in

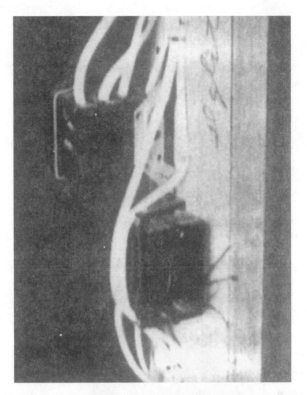

Fig. 9-5 *A nonmetallic switch box, with nails used to mount it. It will hold seven conductors of No. 14 or six No. 12 or No. 10.*

Fig. 9-6 *A nonmetallic switch box, designed with a bracket to mount to the wood or steel studs with SST tools. The box will hold 15 No. 14 conductors, 13 No. 12, or 12 No. 10 conductors.*

Fig. 9-7 *A nonmetallic switch box. It will hold four devices. The bracket will fit between two studs located on 16-inch centers. The box will hold 20 No. 14 conductors, 17 No. 12, or 15 No. 10 conductors.*

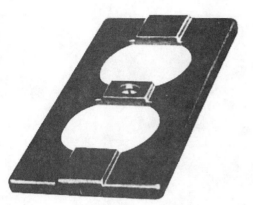

Fig. 9-8 A cover for an outlet box or handy box for a duplex receptacle, nonmetallic.

Fig. 9-9 A nonmetallic cover for a toggle switch.

Fig. 9-10 A nonmetallic surface box. It will hold only three No. 14 conductors and two No. 12 or No. 10 wires.

Fig. 9-11 A nonmetallic conduit box. Round, with four ¹/₂-inch threaded knockouts. It is suitable for fixture mounting. It has 12.5 cubic inches of space and a threaded knockout. It is also available with ³/₄-inch threaded knockouts.

Fig. 9-12 A duplex receptacle cover (REC), with gasket and stainless steel screws. It may also be obtained with a single receptacle hole. Note the REC on the outside to identify the outlet.

Fig. 9-13 A ground continuity tester for use on energized circuits only. It is equipped with a ground and a neutral blade. If the light comes on when it is plugged into a grounding-type receptacle, the ground continuity is complete. If the light does not come on, there is a fault in the circuit. It uses a penlight cell for power.

Fig. 9-14 A box finder and cable tracer. It is used to locate boxes or cables that are hidden from view. To use, plug it into one of the receptacles on the covered box circuit, or attach it to one of the circuit wires. The transmitter will produce a signal. Tune in a small transistor radio to pick up the transmitted signal. Follow the cable by checking sound buildup or fade. As the box is approached, sound builds up due to more wire being folded back into the box. The box will be located at maximum signal strength. The unit is more efficient if the circuit ground wire is disconnected. It is powered by a 9-volt battery.

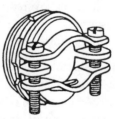

Fig. 9-15 A box connector for entrance service, for nonmetallic sheath cable or nonmetallic flexible tubing. The connector does not make a weatherproof connection to the box.

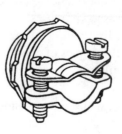

Fig. 9-16 A Romex box connector designed for nonmetallic sheathed cable (Romex) or nonmetallic flexible tubing. It will accommodate 14/2, 14/3, 12/2, 12/3, or 10/2 cable. Romex for home use is usually 14/2 WG or 12/2 WG. Ground wire is assumed to be there, but the designation will usually be 14/2 WG. (WG is the abbreviation for with ground.)

Fig. 9-17 *A BX box connector. Armored cable, or ³⁄₈-inch flexible steel conduit, may be accommodated with this connector. It can handle 14/3, 14/2, 12/2, 12/4, or ¹⁄₂-inch cable.*

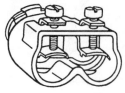

Fig. 9-18 *A BX connector. A duplex connector for armored cable and for ³⁄₈-inch flexible tubing.*

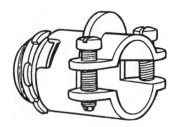

Fig. 9-19 *A BX box connector, with screws for ¹⁄₂-inch flexible steel conduit.*

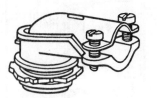

Fig. 9-20 *A 90° BX box connector, used also for ¹⁄₂-inch flexible steel conduit. It will handle 8/2, 8/3, and 8/2 WG cable.*

the box for a positive holding action. Figure 9-21 shows a fiber bushing inserted in a BX.

Figure 9-22 shows how a *connector* is used to hold the Romex in place. *Note how the grounding clip is attached to the metal box.* Note also that part of the Romex insulation is extended through the connector into the box.

A number of sizes and configurations are available to the electrician for use as receptacles. A sampling of the different configurations available for installation in switch boxes and outlet boxes, both metallic and non-metallic, is shown in Fig. 9-23.

Figures 9-24 through 9-35 (Fig. 9-35 on page 144) show the different types of boxes available for use as switch boxes. Some installation procedures are also depicted.

A handy box or utility box These are designed to be used with either thin-wall or rigid conduit. They can handle a switch or an outlet. If the knockout is removed and the Romex cable is properly protected, a box may be used on exposed surfaces if a connector to hold the Romex inside the box is added (Fig. 9-36).

Figure 9-37 shows the following procedures. Figure 9-37A shows how to connect the existing third wire (green insulated or bare copper) to the green grounding screw on a receptacle; Fig. 9-37B shows how the ground wire from the green grounding screw should be connected to a water pipe if no ground is available in the existing system; and Fig. 9-37C shows that the outlet is automatically grounded if properly grounded armored cable (BX or EMT) is used throughout the system. The

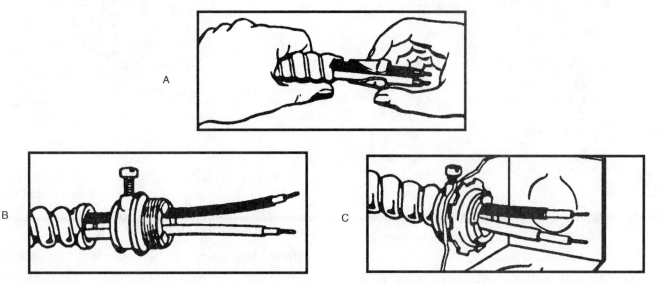

Fig. 9-21 *(A) Bushing inserted into the BX to prevent damage to the insulation of the wire. (B) BX connector is slipped into place over the bushing and metal armor, and the screw is tightened to hold the connector in place. (C) BX connector is held in place through a knockout in the box, by use of a locknut.*

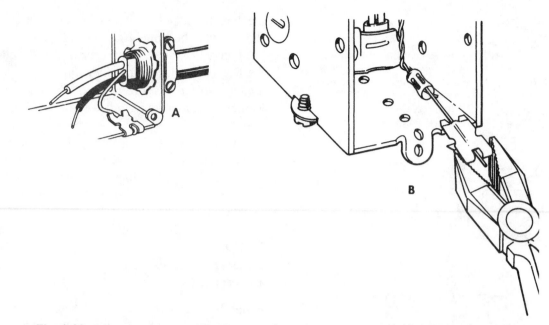

Fig. 9-22 *A Romex connector attached to a box. Note the ground clip for the uninsulated grounded wire.*

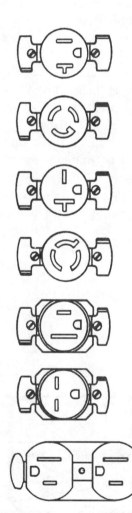

20 A, 125 V,
Three wire,
grounded, two
pole

15 A, 125 V,
Three wire,
grounded, two
pole

20 A, 250 V,
Three wire,
grounded, two
pole

15 A, 250 V,
Three wire,
grounded, two
pole

15 A, 125 V,
Three wire,
grounded, two
pole

15 A, 250 V,
Three wire,
grounded, two
pole

15 A, 125 V,
Three wire,
grounded, two
pole

Fig. 9-23 *Single and duplex receptacles for various current ratings and voltages.*

Fig. 9-24 *A switch box for Romex. It can be mounted without nails. The bracket on the side has an "eagle claw" so that the tabs may be hit with a hammer and driven into the stud without using nails. Note the beveled corner and clamps for the Romex.*

grounded outlet provides protection against shock hazards, such as defective internal wiring on a hand drill.

Romex is protected in an exposed position against a wall in a basement (Fig. 9-38).

Handy-box extension   Sometimes this is added if the proper volume is not available for the necessary number of conductors making their entrance and exit

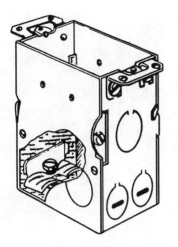

Fig. 9-25 A square-corner Romex switch box, with clamps. This one is deeper than the standard size.

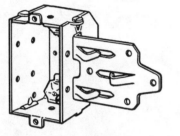

Fig. 9-26 This beveled-corner switch box for Romex may be mounted without nails.

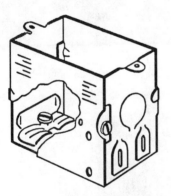

Fig. 9-27 A square-corner switch box with Romex clamps inside. This one does not have plaster ears, but has side leveling ridges and tapped grounding holes.

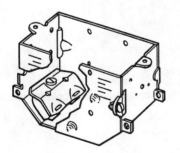

Fig. 9-28 A bevel-corner Romex switch box with tapped rounding holes and side leveling ridges. Nail holes are located on the brackets outside the box. A Romex clamp is included.

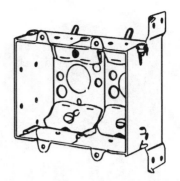

Fig. 9-29 Two switch boxes, ganged. They have Romex clamps and claws, and nails. The claws may be hit with a hammer to hold the box in place until the nails can be inserted.

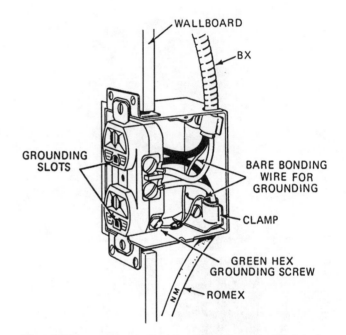

Fig. 9-30 Note how the bare bonding wire is attached to the green hex ground screw on the receptacle. This view also shows the location of clamps inside the box to hold both the BX and the Romex.

Fig. 9-31 A square-corner Romex switch box without plaster ears, but with side leveling ridges and the side nail capability.

Fig. 9-33 *A beveled-corner switch box mounted with two nails.*

Fig. 9-34 *A front view of a mounted beveled-corner switch box.*

Fig. 9-32 *A switch box with nailing holes and a hammer driving the nails into the stud.*

from the utility box. An extension simply slips onto the screws of the utility box and is tightened into place. A cover plate or switch is then placed in the box, or over the extension (Fig. 9-39).

Fig. 9-35 *An insulated wire grounded to the switch box by using the tapped grounding screw hole provided. The insert shows the wire attached to the box by a screw.*

Fig. 9-36 A handy box (utility box) with or without nailing brackets for thin-wall or rigid conduit.

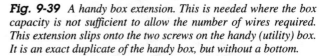

Fig. 9-39 A handy box extension. This is needed where the box capacity is not sufficient to allow the number of wires required. This extension slips onto the two screws on the handy (utility) box. It is an exact duplicate of the handy box, but without a bottom.

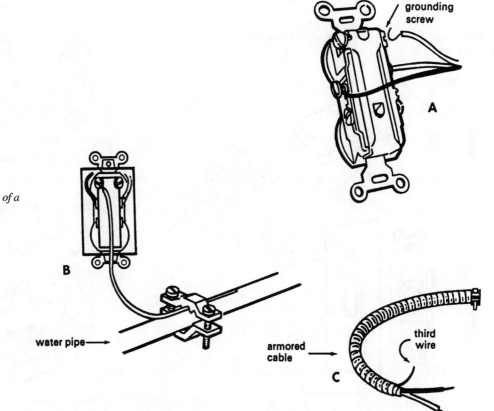

Fig. 9-37 Grounding of a receptacle.

green grounding screw

A

B

water pipe →

armored cable →

third wire

C

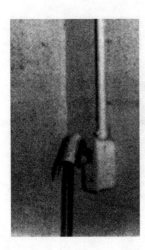

Fig. 9-38 A handy box being used with ½-inch thin-wall to protect the Romex from damage. This type of installation is usually found in a home in the basement where there is the possibility of damage to the unprotected Romex by the movement of heavy objects.

Cover plates   Because utility boxes must be totally enclosed, several plates are available to fit individual situations. Figures 9-40 through 9-43 show the types available.

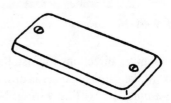

Fig. 9-40 A blank cover for the handy box.

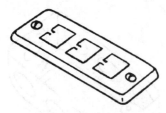

Fig. 9-41 *A handy box cover with three knockouts. Captive screws and a galvanized finish make up the total cover. Captive screws will not fall out when unscrewed from a handy box. They must be intentionally unscrewed to remove them from the cover.*

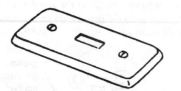

Fig. 9-42 *A handy box switch box cover for a tumbler or toggle switch with square handles.*

Fig. 9-43 *A handy box cover for a 20-ampere, single-receptacle outlet.*

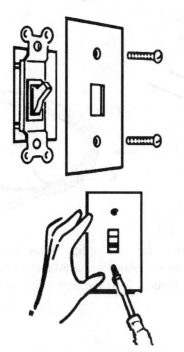

Fig. 9-44 *How to install a switch plate.*

Figure 9-44 gives directions for installing a switch plate on a box. Make sure the power is turned off if you are installing a plate on existing wiring. Note the different types of plates: double-switch plate, combination switch plate and double-outlet plate, double-outlet plate, and switch plate and single-outlet plate.

Octagonal boxes of 3- and 4-inch sizes are available for use with Romex, BX, thin-wall, or rigid conduit. They come with or without clamps and have knockouts of various sizes. All of them have two screws for attaching a cover plate. Extensions are available for increasing the space to hold a number of conductors (Figs. 9-45 through 9-49).

Sometimes it is necessary to place a utility box or outlet box between two studs or joists. For this situation, an expandable bar hanger comes in handy. It expands to fit between studs or joists centered on 14, 16, 18,

Fig. 9-45 *A 4-inch octagonal box for Romex or BX. Note the tapped grounding screw holes. Clamps are included. This type is not to be used in ceilings without mud rings.*

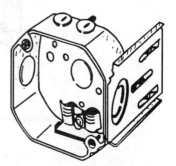

Fig. 9-46 *A 4-inch octagonal outlet box for Romex and BX with bracket. No nails are needed for mounting. This one includes clamps and has a tapped grounding screw.*

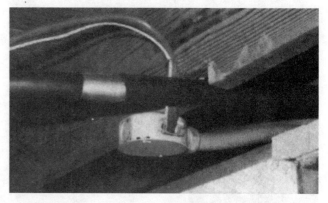

Fig. 9-47 *An outlet box used as a junction box for Romex and flexible metal conduit. The cover is blank.*

Fig. 9-48 A fixture hanger with safety clips.

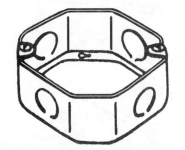

Fig. 9-49 A 3-inch octagonal-box extension ring.

and 20 inches. The center clip is adjustable to any location between the maximum and minimum setting (Fig. 9-50). The bar hanger in Fig. 9-51 has nail holes for mounting between studs or joists. The sliding clip facilitates easy location of the box. Figure 9-52 shows a Romex bar set with an outlet box. It will expand to fit joists on 14-, 16-, 18-, or 20-inch centers.

Fig. 9-50 An expandable bar hanger. It can be adjusted for 14-, 16-, 18-, and 20-inch centers.

Fig. 9-51 A bar hanger, 18 inches long, with sliding clip.

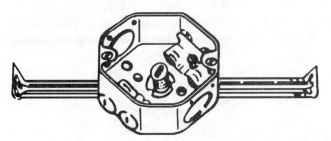

Fig. 9-52 An outlet box mounted on an expandable bar hanger.

Occasionally, a cover is needed that will extend a box to fit flush with the wall. A raised cover (⅝-inch) for a 4-inch octagonal outlet box is shown in Fig. 9-53. It is also available in other sizes.

Sometimes an outlet with a steel cover plate is required for areas subject to possible mechanical damage of the plate. A steel cover plate for a 4-inch octagonal box will accommodate a duplex receptacle (Fig. 9-54). Figure 9-55 shows a blank cover plate for a 4-inch octagonal box. However, a blank box cover is not always the answer, especially if you want to mount an object from the box. Thus, a blank cover with a ½-inch knockout might be utilized, so that a fixture may be mounted from the 4-inch octagonal box it covers (Fig. 9-56).

Fig. 9-53 A 4-inch octagonal outlet box with a ⅝-inch raised cover.

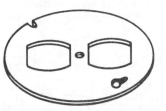

Fig. 9-54 A 4-inch octagonal outlet box cover with flush mounting of a duplex receptacle.

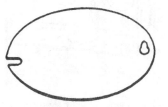

Fig. 9-55 A 4-inch octagonal outlet box cover, blank.

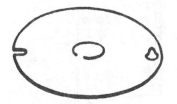

Fig. 9-56 A 4-inch octagonal outlet box cover with a ½-inch knockout.

In Figs. 9-57 through 9-61 you will find an assortment of boxes 4 inches square, some with tapped, grounding-screw holes and others with clamps. Some of these boxes have no brackets for mounting; others have no-nail brackets and self-alignment brackets. Some use nails for mounting. Others will be used on conduit or thin-wall by removing the knockout and inserting the proper connector. There is also an extension for these boxes. A box that is $4^{11}/_{64}$ inches square has a greater cubic inch volume for additional conductors. It is difficult at first to recognize the difference that only $^{11}/_{64}$ inch makes. Once you are working with both the 4- and $4^{11}/_{64}$-inch sizes, however, you will become very accustomed to the difference.

Fig. 9-57 *A 4-inch square drawn box with tapped grounding screw holes and two screws for cover mounting. This one is made for use on rigid conduit or thin-wall conduit, but can also be used for Romex.*

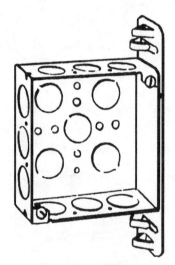

Fig. 9-58 *A 4-inch square box for thin-wall or rigid conduit. It can also be used for Romex if the proper connectors are used.*

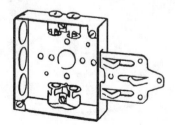

Fig. 9-59 *This 4-inch square box has a bracket for mounting against the stud without nails.*

Covers for a 4-inch square box are many and varied. Figures 9-62 through 9-66 show various types and configurations. Note figure captions to obtain more details of size, function, and shape.

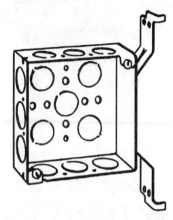

Fig. 9-60 *A 4-inch square box for thin-wall, Romex, BX, or rigid conduit.*

Fig. 9-61 *A 4-inch square box extension. Note how it fits over an existing box.*

Fig. 9-62 *A ¹/₂-inch square box. This one will accommodate one duplex flush receptacle.*

Fig. 9-63 *A square cover with toggle switch and duplex holes.*

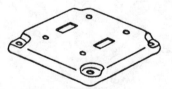

Fig. 9-64 *A switch cover for a 4-inch square box.*

Fig. 9-65 *A ¼-inch raised single-device cover for a 4-inch square box. It is used where a larger box is needed to hold the wires, but a single receptacle or switch is mounted in the box.*

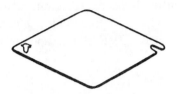

Fig. 9-66 *A flat cover blank is used on a 4-inch square box.*

FIXTURES AND SPLICES
Lighting Fixtures

Lighting fixtures may be installed in several ways. There are hanger supports that thread onto a threaded stud, mounted in the box, as in Fig. 9-67. Straps and machine screws are used to mount the fixture in Fig. 9-68. If there is no stud, the metal strap may be used to hold the canopy of the light fixture in place, as in Fig. 9-69.

Fig. 9-67 *Mount large drop fixtures by simply using a screw hanger support onto the threaded stud in the outlet box. Use solderless connectors (wire nuts) to connect the electrical wires and a grounding clip for the extra uninsulated ground wire. Raise the canopy and anchor in position by means of a locknut.*

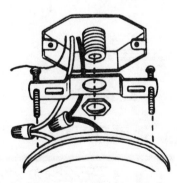

Fig. 9-68 *Outlet box has a stud in this case. Insert the machine screws in threaded holes of the metal strap shown. Slip the center hole of the strap over the stud in the outlet box. Hold the strap in position by a locknut. Connect wires with wire nuts, and slip the canopy over the machine screws; fit flush and secure the fixture with two cap nuts. Don't forget to anchor the uninsulated ground wire to the box with a ground clip or to the box's threaded grounding screw.*

Glass-enclosed ceiling fixtures are easily attached to the ceiling box by using a threaded stud (Fig. 9-70). Wall lights may be installed by using the same method. Make sure the outlet found in the wall fixture is wired for full-time service—not controlled by the switch that turns the light on and off (Fig. 9-71).

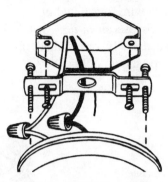

Fig. 9-69 *If there is no stud, insert the machine screws as shown here. Fasten the ears of the outlet box and the strap with screws. Then align the canopy onto the two screws pointing down and cap off with cap screws.*

Fig. 9-70 *Glass-enclosed fixtures can be installed by the method shown here.*

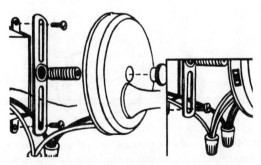

Fig. 9-71 *Wall brackets or lights are installed by strapping to the ears of the box, then using a nipple and cap to complete. Don't forget the ground wire and the ground clip. If there is an outlet which is "always on," wire according to the insert.*

Replacing a Damaged Lamp Cord

Plugs and cords sometimes need repair. Replacement plugs with a heavy-duty capability should be used. Figure 9-72 shows how the *underwriter's knot* is used to relieve strain put on the connection when the cord is pulled. The two plugs are replacement types that do not have a third wire for grounding.

Some lamps use an *in-line switch* for control. It may be replaced easily, or installed, if needed, by following the steps shown in Fig. 9-73.

An adapter is available for three-pronged plugs to make them usable on the older two-wire systems installed in most homes before 1960 (Fig. 9-74). The lug is inserted under the screw that holds the cover plate onto the outlet.

Splicing Wires

Manufacturers of a specific type of fastener or clip have specialized methods that they recommend to splice wire. Most of these will be covered in Figs. 9-75 through 9-93 (Fig. 9-93 on page 153). Read the captions under the fig-

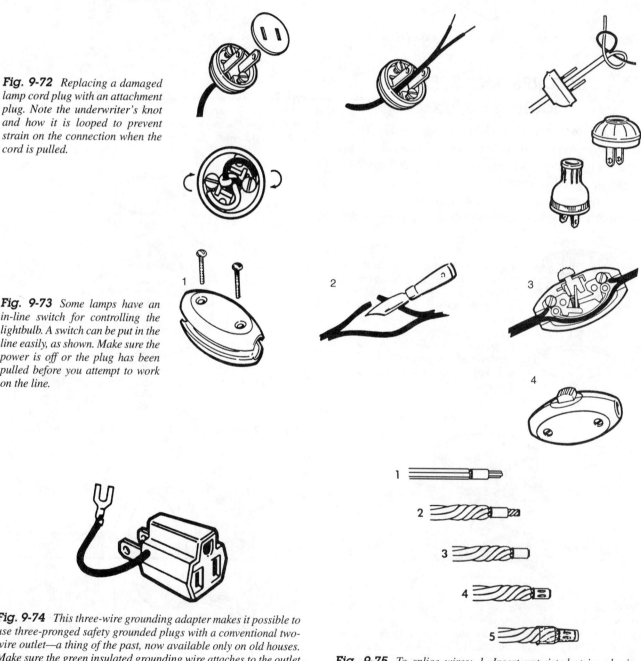

Fig. 9-72 *Replacing a damaged lamp cord plug with an attachment plug. Note the underwriter's knot and how it is looped to prevent strain on the connection when the cord is pulled.*

Fig. 9-73 *Some lamps have an in-line switch for controlling the lightbulb. A switch can be put in the line easily, as shown. Make sure the power is off or the plug has been pulled before you attempt to work on the line.*

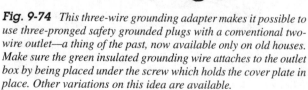

Fig. 9-74 *This three-wire grounding adapter makes it possible to use three-pronged safety grounded plugs with a conventional two-wire outlet—a thing of the past, now available only on old houses. Make sure the green insulated grounding wire attaches to the outlet box by being placed under the screw which holds the cover plate in place. Other variations on this idea are available.*

Fig. 9-75 *To splice wires: 1. Insert untwisted stripped wires through the splice cap. 2. Twist wires. 3. Cut wires flush with cap. 4. Insert tool—squeeze to crimp. 5. Snap on a nylon insulator.*

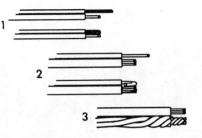

Fig. 9-76 *Splicing solid and stranded wires: 1. Loop stranded wire, if No. 16 or smaller. 2. Loop solid wire, if smaller than stranded wire. 3. When you join two or more solid wires to a larger stranded wire, twist the solid wires together.*

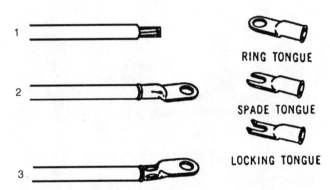

Fig. 9-77 *To insulate splices, just snap on the nylon insulator over the installed splice cap.*

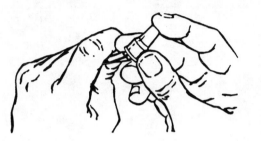

RING TONGUE

SPADE TONGUE

LOCKING TONGUE

Fig. 9-78 *To terminate wires with lugs: 1. Strip wire(s) approximately 5/16 inch. 2. Slip lug on wire with tool in position. 3. Insert tool with flat side of the lug in the up position so that the tongue enters the slot in the latch. Squeeze the tool to crimp.*

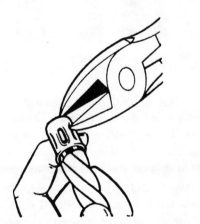

Fig. 9-79 *To remove splicing caps or lugs without damaging the wires, snip splice cap of both ends of one crimp, counteracting cutting pressure with the index finger. Then peel off the cap. To remove the lugs, snip off tongue and proceed as for the splice caps.*

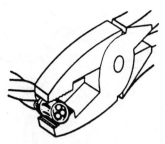

Fig. 9-80 *To remove "cold working" cap, apply pressure alternately at two points 90° apart and between crimps—then pull off the cap. This is most effective when the cap is not full of wire.*

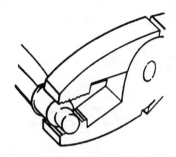

Fig. 9-81 *To remove nylon insulators, apply pressure alternately at two points 90° apart to "cold-work" the metallic retainer—or cut up side of insulator body.*

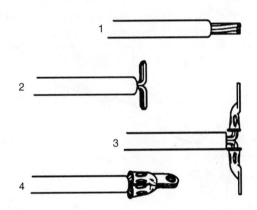

Fig. 9-82 *To terminate No. 6 wire using two lugs: 1. Strip wire approximately 3/8 inch. 2. Untwist wire lay and separate strands into two approximately equal groups. 3. Crimp a lug on each group. 4. Bring flat sides of lugs together.*

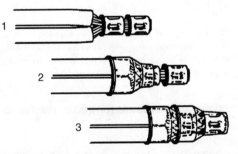

Fig. 9-83 *How to make a strain relief splice for service entrance. 1. Strip both wires approximately 1 1/2 inches and untwist wire lay; install two splice caps, leaving about 1/16 inch between the two caps. 2. Cut off the top half of the tip of an insulator and push remainder over the cap. 3. Snap another insulator over the end cap.*

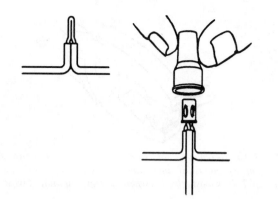

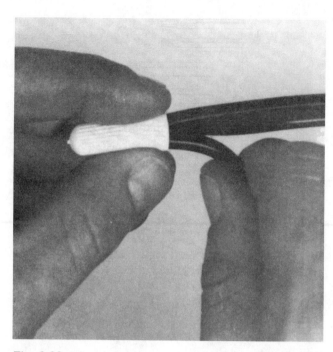

Fig. 9-84 *Tee-tap (where slack permits): 1. Strip and loop wire. 2. Add stripped tap wire, splice, and snap on the insulator.*

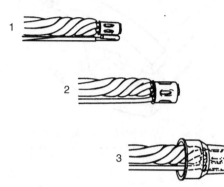

Fig. 9-85 *Adding wire (No. 12 or smaller): 1. Loop wire to be added and press into the crimped indentation of the cap. 2. Force the splice cap over the joint, then crimp. 3. Insulate with a snap-on insulator.*

Fig. 9-88 *Wire nut insulator used to splice wires inside a switch box or outlet box.*

Fig. 9-89 *Wire nut insulators used in various sizes for different wire sizes. They are also available in color for coding of wires.*

Fig. 9-86 *Grounding connection. Splice wires with a splice cap, leaving one wire extending through the cap to permit attachment of a lug.*

Fig. 9-90 *Wire nuts with a Bakelite case and copper inside.*

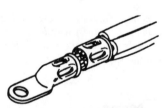

Fig. 9-87 *Terminating two or more wires with a lug. This is a technique used if the barrel capacity of the lug is too small to hold the wires. Splice wires, leaving one or more wires extending through the cap to permit attachment of the lug as shown.*

ures to learn the proper procedure for making and insulating a splice.

Mounting Boxes in Existing Walls

A switch box may be mounted by using the existing wall as part of the support device (Figs. 9-94 and 9-95).

A special clip that will exert pressure from the inside of the wall when the screws are tightened is shown in Fig. 9-94. In Fig. 9-95 a piece of sheet metal is inserted to fold over inside the box and fit behind the dry wall to support the box. Note that the plaster ears on the box are important. They add to the support when a switch is turned on and off or when a plug is pulled from an outlet mounted in the box. These are two of the many methods used to add an outlet to an existing wiring system.

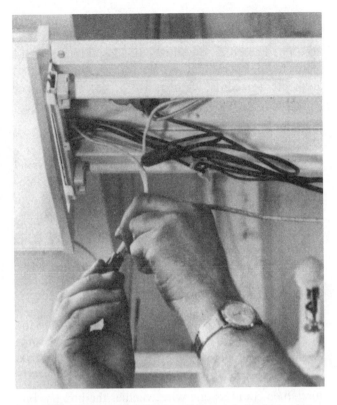

Fig. 9-91 *Using wire nuts for the splicing needed in the insulation of fluorescent fixtures.*

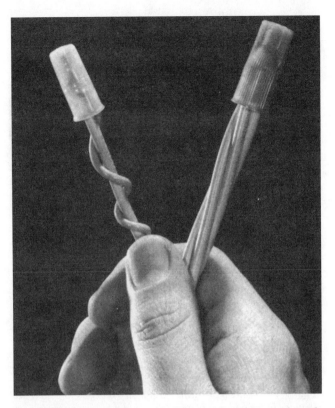

Fig. 9-92 *Two- and three-wire splices with plastic see-through insulators.*

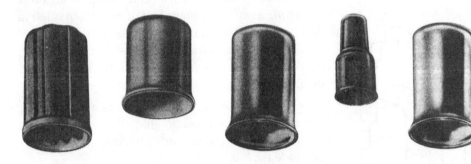

Fig. 9-93 *Various sizes and shapes of end caps for splicing wires.*

NUMBER OF CONDUCTORS IN A BOX

Every box used for switches, receptacles, or fixtures has a volume. This volume determines the number of wires that can be brought into the box safely. The National Electrical Code has a table of sizes and the maximum number of wires (conductors) allowed in a box. This is one of the most important things you must consider when you are wiring a house. The inspector looks here to make sure you have done the job properly.

Table 9-1 shows the number of wires in a box. This assumes all the wires are the *same* size. If the wires are not the same size, you will have to check the volume of the box and then take a look at Table 9-2. Here, the volume required for each wire is shown. Just add the number of wires of a particular size and multiply by the volume required by that conductor. Then do the same thing for the next size conductor. Once you have figured all the various sizes and their volume requirements, just add them. This will tell you whether you have too much for the volume of the box.

Examples   A 4- × 1¼-inch round or octagonal box will have a minimum cubic inch capacity of 12.5. If you use all No. 14 wires, it can handle six of them. If you use No. 12 wires, it can hold only five wires. However, if you mix the two sizes, you have to go to Table 9-2 to see the volume requirement of each conductor. If you look at the table, you will see that the No. 14 wires must have 2.0 cubic inches for each conductor. The No. 12 wires will need 2.25 cubic inches for each conductor.

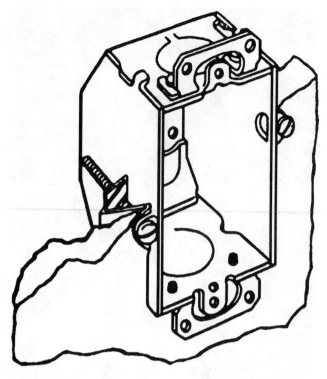

Fig. 9-94 *Using a box with "grip-tight" brackets to hold a box in place in the addition of a switch or outlet to existing wiring.*

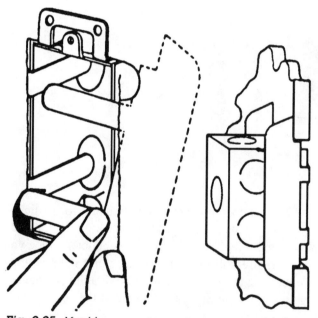

Fig. 9-95 *Metal box supports can ensure a stronger job. Insert supports on each side of the box. Work supports up and down until they fit firmly against inside surface of the wall. Bend the projecting ears so that they fit around the box.*

You have only 12.5 cubic inches in the box. Therefore, you can have only three wires of No. 12 for 6.75 cubic inches and two wires of No. 14 at 4 cubic inches total. Add them up and you get 10.75 cubic inches. If you

use three of No. 14 and three of No. 12, you get 12.75 cubic inches, or 0.25 cubic inch more than the box has to offer. Therefore, the three No. 12 and three No. 14 wires are not permitted in this size box. You could go to the next-size box, which is a 4- × 1½-inch round or octagonal. Such a box has 15.5-cubic-inch capacity.

Which Wires to Count

Ground wires   No matter how many ground wires come into a box, a deduction of only one conductor must be made from the number of wires shown in Table 9-1. Any wire that runs unbroken through a box is counted as one wire. The ground wires may be in nonmetallic cable or ground wires that are run in metal or nonmetallic raceways.

Spliced wires   Any wire that comes into a box and is spliced is counted as one wire. It may be crimped or twisted in its connection. If a wire that comes into the box is connected to a wiring device terminal, it is counted as one wire.

Clamps in boxes   Cable clamps, hickeys, and fixture studs count as one wire whether the box has one clamp, two clamps, or any combination of clamps, studs, or hickeys.

You may remove unused cable clamps from a box to provide more space in the box. If one clamp is left in the box, you must count the clamp as one conductor. If both clamps are removed and you use a box connector, then the clamp is not deducted from the total allowed in the box.

Grounding jumper   If a jumper is used from the box screw to the receptacle grounding terminal, then the jumper is not counted as a conductor since it does not leave the box.

If a switch or receptacle has a grounding strip on it, this must also be counted as a conductor in terms of the number of conductors used in a box.

Mounting boxes   One of the factors involved in mounting boxes and electrical equipment is the prevention of fire. Also, in case of fire, it must not be allowed to move from one level to another by way of a raceway or hole. The use of a hexagonal or square box in ceilings where there is sheetrock is not permitted. If they are used, they must have a "mud ring" installed to prevent fire from being allowed to contact wall or ceiling materials that may burn.

Single-gang boxes   In single-gang boxes, the nonmetallic cable does not have to be clamped to the box. It should be secured to a stud within 8 inches of the box, however (Fig. 9-96).

Table 9-1 Boxes and Maximum Number of Wires*

Box Dimension (Inches), Trade Size or Type	Minimum capacity, cu in	Maximum Number of Conductors			
		No. 14	No. 12	No. 10	No. 8
4 × 1¼ Round or octagonal	12.5	6	5	5	4
4 × 1½ Round or octagonal	15.5	7	6	6	5
4 × 2⅛ Round or octagonal	21.5	10	9	8	7
4 × 1¼ Square	18.0	9	8	7	6
4 × 1½ Square	21.0	10	9	8	7
4 × 2⅛ Square	30.3	15	13	12	10
4¹¹/₁₆ × 1¼ Square	25.5	12	11	10	8
4¹¹/₁₆ × 1½ Square	29.5	14	13	11	9
4¹¹/₁₆ × 2⅛ Square	42.0	21	18	16	14
3 × 2 × 1½ Device	7.5	3	3	3	2
3 × 2 × 2 Device	10.0	5	4	4	3
3 × 2 × 2¼ Device	10.5	5	4	4	3
3 × 2 × 2½ Device	12.5	6	5	5	4
3 × 2 × 2¾ Device	14.0	7	6	5	4
3 × 2 × 3½ Device	18.0	9	8	7	6
4 × 2⅛ × 1½ Device	10.3	5	4	4	3
4 × 2⅛ × 1⅞ Device	13.0	6	5	5	4
4 × 2⅛ × 2⅛ Device	14.5	7	6	5	4
3¾ × 2 × 2½ Masonry box/gang	14.0	7	6	5	4
3¾ × 2 × 3½ Masonry box/gang	21.0	10	9	8	7
†FS—Minimum internal depth 1¾ single cover/gang	13.5	6	6	5	4
FD—Minimum internal depth 2⅜ single cover/gang	18.0	9	8	7	6
FS—Minimum internal depth 1¾ multiple cover/gang	18.0	9	8	7	6
FD—Minimum internal depth 2⅜ multiple cover/gang	24.0	12	10	9	8

Table 9-2 Volume Required for Each Wire Size*

Size of Conductor	Free Space within Box for Each Conductor
No. 14	2.00 cubic inches
No. 12	2.25 cubic inches
No. 10	2.50 cubic inches
No. 8	3.00 cubic inches
No. 6	5.00 cubic inches

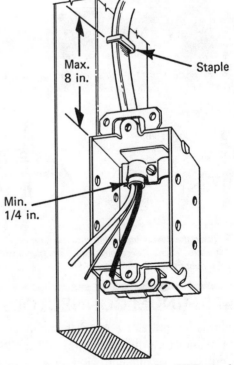

Fig. 9-96 *Installation of the nonmetallic (NM) wire known as Romex in a single-gang box.*

If the box is round, or square two-gang or three-gang, the cable must be clamped to the box (Fig. 9-97).

Installing a connector Figure 9-98 shows the newer type of Romex box connector. There are no screws needed. Insert the Romex cable as shown in Fig. 9-99. Slide the connector over the cable from inside the box. Then squeeze the connector on the outside of the box to make a fit with no movement of the wire or connector.

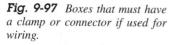
Fig. 9-97 Boxes that must have a clamp or connector if used for wiring.

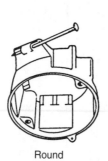
Round

Square—Two-Gang

Three-Gang

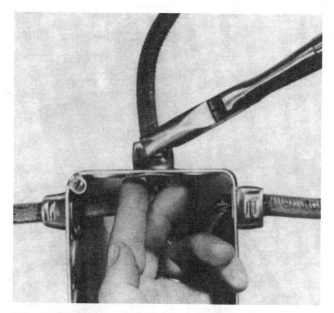
Fig. 9-98 Romex connectors—no screws, no locknut. Just use a pair of pliers to exert pressure to hold the Romex in place.

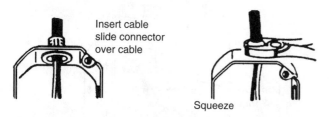
Insert cable slide connector over cable

Squeeze

Fig. 9-99 Insert the Romex cable; slide the connector over the cable from inside the box. Then squeeze the connector on the outside of the box to make a tight fit with no movement of the wire or connector.

ALUMINUM CONNECTORS

Aluminum is a more active metal than copper. In the presence of moisture, the aluminum will erode. This susceptibility to galvanic corrosion, or electrolysis, is a problem because it will weaken an aluminum connection (Fig. 9-100).

The electrolytic action between aluminum and copper can be controlled by plating the aluminum with a neutral metal (usually tin). Such plating prevents electrolysis from taking place. It also helps keep the joint tight. As an additional precaution, however, a joint-sealing compound should be used. The compound contains fine zinc particles which break through the oxide film that forms on an aluminum connector. It also seals out air and moisture.

Upon exposure to air, aluminum immediately becomes coated with a film of oxide. This oxide film has the properties of a ceramic, therefore insulating the aluminum and increasing joint resistance. The film must be removed, or penetrated, before a reliable aluminum joint can be made.

Aluminum connectors are designed to bite through the film of oxide on the aluminum as the connector is applied to the conductors. It is further recommended that the conductor be wire-brushed and preferably coated with a joint compound to guarantee a reliable joint.

Aluminum Wire for Home Wiring

Aluminum wiring came into use when there was a shortage of copper wire. Many houses built between 1965 and 1971 used such wiring. The aluminum wire was usually No. 12 instead of the copper No. 14.

Use of aluminum wire can be a safety hazard. The greatest problems have occurred where there are extreme temperature differences. Aluminum expands and contracts at a greater rate than copper. Most of the potential problems can be eliminated by tightening all connections at outlets and switches. Materials now available can ensure that copper and aluminum can be safely connected with wire nuts or by other means.

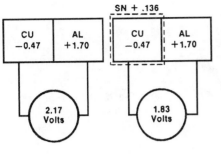

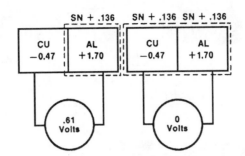

Electrode potential. (Values show only the relative activity of metals. Activity decreases in going from copper to aluminum, tin-plated copper to aluminum, copper to tin-plated aluminum, tin-plated copper to tin-plated aluminum.) Cu = copper, Al = aluminum, Sn = tin.

Fig. 9-100 *Electrode potential.*

There are two types of aluminum wire in general use today: all-aluminum and copper-clad aluminum. Bonded to the surface of the copper-clad aluminum is a coating of copper. This eliminates some of the problems that occurred when copper and aluminum were placed together in a wire nut or some other connector.

When aluminum wire was first introduced, it was used just as copper wire. This created some problems, since the aluminum wire was clamped in a brass or copper terminal screw or connector. The heat created by current flowing in the wire causes the aluminum to *cold-flow* out of the connector. When the connection cools, it becomes a little loose. As the current is turned on and off, the heating and cooling cause still more aluminum to become extruded. Eventually the wire can heat up and cause a fire.

In damp areas this can cause an electrolytic action to take place. Copper or brass terminal screws react with the aluminum.

Not all switches and receptacles that were intended for 15- or 20-ampere service are marked. However, those intended for use with copper (CU) and aluminum (AL) are marked CU-AL. Recently, the UL changed its rating of 15- and 20-ampere switches and receptacles and introduced the CU-ALR designation for aluminum in the 15- and 20-ampere rating. The *R* stands for *revised*. The composition of the aluminum wire has been changed also. There are special chemical and physical properties specified by the UL before the wire gets approval.

Devices marked CU-AL should be used with copper wire only.

Devices marked with CU-ALR (or CO-ALR) may be installed with aluminum, copper-clad aluminum, or copper wire. Devices with a push-in connection must *not* be used with all-aluminum. They can be used on copper-clad aluminum or copper wire.

Devices rated 30 amperes or more are *not* marked with CU-ALR or CO-ALR. Those not marked, and rated at 30 amperes or more, can be used with copper-clad aluminum or copper wire.

CIRCUIT PROTECTIVE DEVICES

The circuit breaker and the fuse are the circuit protective devices most frequently used to ensure that a circuit operates within its designed limits.

Fuse

The fuse is a device that destroys itself when it operates. That is, a fusible link is destroyed when it performs its designed function. An overcurrent will cause a piece of metal in the fuse to heat up and finally melt, thereby interrupting the circuit. This occurs because the fuse is inserted in series with the device it is to control.

One type of fuse is shown in Fig. 9-101. It is a spring-loaded one that can disrupt the circuit if a unit overheats and is removed from its contact, as in D, or if it burns or melts, as in C. Part A of Fig. 9-101 shows the unit in its complete form, with a screw-in base for insertion in a fuse box. A cutaway view with the construction of this type of fuse is shown in B. These are normally found in outmoded wiring in old homes.

Circuit Breakers

A circuit breaker is a device that, after breaking a circuit, can be reset by turning the handle to *off* and then to *on* (Fig. 9-102).

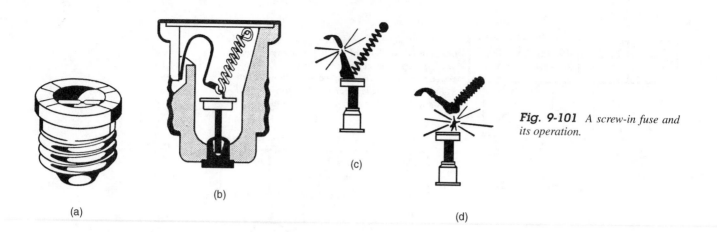

(a)

(b)

(c)

(d)

Fig. 9-101 *A screw-in fuse and its operation.*

Fig. 9-102 *A cutaway view of a circuit breaker.*

proper size is used to hold or clamp the wire so it will not be moved easily by an external force.

Fig. 9-103 *A single 20-ampere circuit breaker.*

The circuit breaker is mounted in a distribution box by snapping it into place. The hot wire (black or red) is attached by inserting it under the screw and tightening. A knockout blank in the distribution box must be removed, to allow for the handle and top of the circuit breaker to be exposed. Figure 9-103 shows a circuit breaker for 20 amperes and a single connection.

Another type is the 100-ampere double circuit breaker, with the two breakers tied together, or a common trip for a two-pole arrangement (Fig. 9-104). It requires two spaces in a distribution box.

Circuit-breaker box The circuit breaker must be housed to protect it from abuse. It also has exposed areas that could cause shock if a person touched them when the circuit was live. The opening for the cable to be fed into the box is easy to accomplish, as shown in Fig. 9-105. The knockout is removed by using a screwdriver and a pair of pliers. Then the connector of the

Fig. 9-104 *A double circuit breaker, 100-ampere, with common trip.*

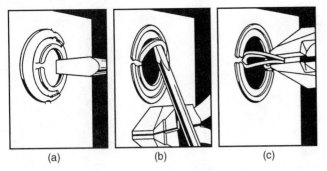

Fig. 9-105 *Removing the knockout from a panel box for circuit breakers.*

Placing the circuit breakers into the panel must be done by removing the slots that allow the circuit breaker to protrude through the box for easy operation. By placing this panel over the circuit breaker, it is possible to protect the operator of the circuit breaker from exposure to 240 volts within the circuit breaker box (Fig. 9-106).

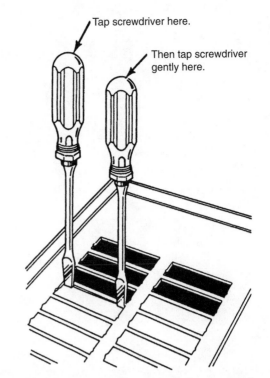

Tap screwdriver here.

Then tap screwdriver gently here.

Fig. 9-106 *Using a screwdriver to take out the slots in a front panel of a circuit breaker box. Note how the screwdriver is tapped on the end first at one position and then at the other. This knocks the slot material loose, and it can be taken out by hand.*

10
CHAPTER

Wiring with BX, EMT, Flexible, Plastic, and Rigid Conduit

THERE ARE A NUMBER OF METHODS USED TO enclose wires for houses. The primary objective is to make sure the electrical wiring and its many points of junction or connection cannot cause a fire under normal operating conditions.

As a rule, conduit is not used for wiring houses. It is of great use in industrial and commercial locations, but becomes rather expensive for use in homes. The metal of the conduit protects the wire insulation and the wires themselves from physical damage that may result in shorts or opens. These shorts may cause overheating and fire. Arcing can cause explosions in some atmospheres. The enclosed conduit type of wiring can be installed to prevent explosions caused by arcing. In some locations, it is necessary to enclose totally a switch or outlet to prevent an explosion. Hospitals, for instance, need special attention paid to the operating rooms. Here, gases used for anesthetizing patients may cause an explosion if ignited by a spark.

Some of the applications of conduit to industrial and commercial locations will be shown here for your information. You may want to add a special wiring arrangement if you have a special-purpose room in a home you are building or remodeling.

BX CABLE

BX cable is the name applied to armored or metal-covered wiring. It is sometimes used in home applications. BX (the letters stand for metal-covered, flexible wire) sometimes meets the need in home applications for flexible wiring. It is used to connect an appliance, such as a garbage disposal unit, which vibrates or moves a great deal. For example, a garbage disposal unit is installed under a sink in the kitchen. The unit will vibrate when in normal use. This vibration calls for a flexible conduit or wire. The BX is attached to a junction box on the wall and through a special connector to the disposal unit.

BX, however, presents some rather unique problems. It must be cut properly. If it is not properly cut, the insulation covering the wires inside may be shorted to the protective metal covering. There is a fiber bushing designed to fit into the cut end of the BX. This bushing insulates the sharp end of the BX (where it was cut) and prevents shorts.

BX Cutter

A tool has recently been designed to cut BX without the ragged edges produced by a hacksaw (Fig. 10-1). It is very efficient and can be carried on an electrician's belt with other tools.

Using the cutter Once the BX cable has been secured in the cutter, with the thumbscrew on the bottom,

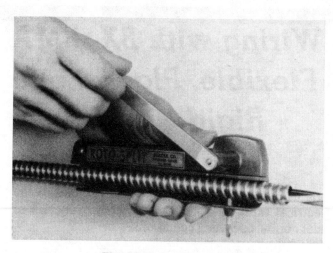

Fig. 10-1 BX cable ripper.

squeeze the handle. Squeezing the handle overcomes the force of the built-in spring. At the same time, it lowers the circular cutter against the BX casing. Do not apply too much pressure if you want a fast cut. At the point when the force required to turn the handle suddenly reduces, even when you increase the squeezing force, you have reached the pressure limit. Stop and release the tool. Release the hand pressure and turn the crank slightly. The tool should snap back to its original position for easy removal from the BX.

After the tool has been removed from the BX, hold the cable on each side of the slit and twist the casing counterclockwise until the casing separates. The insulated wires inside the BX casing are now exposed. If this separation doesn't happen easily, you haven't cut the BX completely. You will therefore have to go back and insert the tool again to finish cutting the armored cable housing.

Once the BX has been cut, you must install it properly in a box for termination. There are a number of connectors for this type of job.

FITTINGS
Connectors

Box connectors These are designed to secure cable, metallic or nonmetallic, to junction enclosures. These connectors are not watertight and are used indoors (Fig. 10-2).

They are available in straight, 45°, and 90° angle styles. Some types have a screw-clamp method of gripping the cable; others utilize a squeezing principle (Figs. 10-3 and 10-4). Their purpose is to provide a clamping action without causing injury to the cable. The clamping action prevents connected wires from getting pulled apart by stress put on the cable from outside the enclosure (Fig. 10-5).

TWO SCREW
CLAMP TYPE

SINGLE SCREW
CLAMP TYPE

SQUEEZE TYPE

ROMEX

SERVICE
ENTRANCE

CORD OR
BARE GROUND

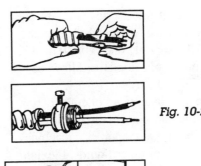

A

B

C

Fig. 10-2 *Box connector clamps.*

90° - 45° FOR
BX CABLE

STRAIGHT FOR
BX CABLE

DUPLEX FOR
BX CABLE

Fig. 10-3 *BX cable connectors.*

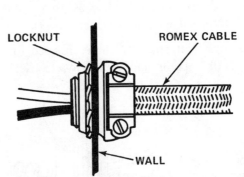

LOCKNUT

ROMEX CABLE

WALL

Fig. 10-5 *A straight, two-screw, clamp-type connector used for Romex. In some cases, this type of connector can be used with BX. Note the difference in Fig. 10-7A.*

Box connectors used with metal-clad cable provide a continuous ground through the cable armor to the metal box. The cable is attached to the box with a connector. This ensures a good ground connection. Box connectors are easy to install. Insert them in a knockout or drilled hole, and secure them in place with a locknut. Many outlet and switch boxes have built-in cable clamps (Fig. 10-6). When such clamps are provided, no box connector is needed.

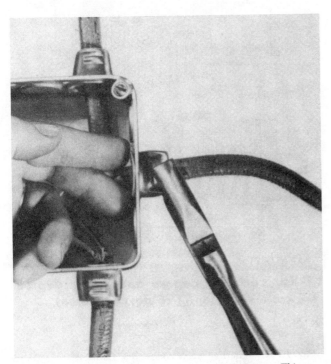

Fig. 10-4 *A squeeze-type connector used on Romex. This type cannot be used on BX.*

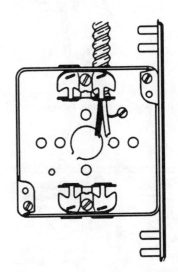

Fig. 10-6 *An outlet box with BX clamps built in.*

Included in the box connector series is a straight duplex connector. This will take two metal-clad cables. All other connectors will take only one cable. The 45° and 90° types have removable hoods to allow easy insertion of the cable (Fig. 10-7).

Box connectors are used in all types of construction where the use of metallic and nonmetallic cables and raceways is permitted. They are also applied to machines that have a light spray of oil or other liquids. Flexible cables are practical in most tight-area wiring (Fig. 10-8).

Ninety-degree knockout box connectors These are designed for use in connecting conduit to outlet boxes. They may be used on other similar devices. Two types are available—threaded and no-thread (Fig. 10-9).

Cable connectors (watertight) These are used for entrance of flexible cords and cables to electrical

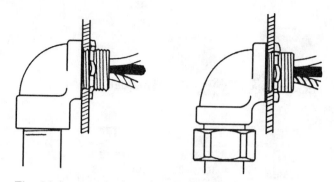

Fig. 10-9 *The 90° knockout elbow. This one has a conduit fitting and attaches to the box with a locknut.*

equipment (Fig. 10-10). They provide clamping action and reduce cable insulation abrasion and wear. Watertights seal out vapor and dirt. They are both watertight and oiltight. Thus, they help prevent deterioration of valuable enclosed mechanisms. Their use can also prevent wire terminal strains. Strains result in dangerous broken connections. Use of watertights can prevent short circuits due to abrasion at the cable connector.

These connectors are made of lightweight aluminum. They consist of four parts: *body, grommet, ring,* and *cap.* An oil-resistant *neoprene* grommet provides a tight seal and firm grip on the flexible cord, or cable. The ring under the clamping cap allows the grommet to be compressed tightly around the flexible cord, or cable, without friction on the cap.

Watertight connectors are available in straight and 90° angles. Hub sizes are $3/8$, $1/2$, $3/4$, 1, and $1 1/4$ inches.

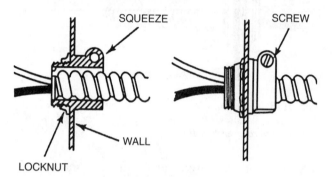

Fig. 10-7A *The 45° and 90° connectors for BX cable.*

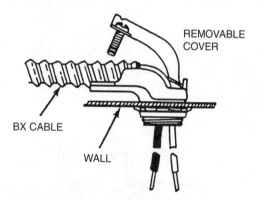

Fig. 10-7B *A box connector with BX clamp. Note the removable cover for ease in placing wires into the box.*

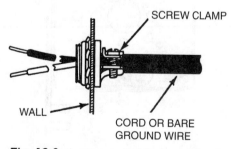

Fig. 10-8 *A connector holding a cord in a box.*

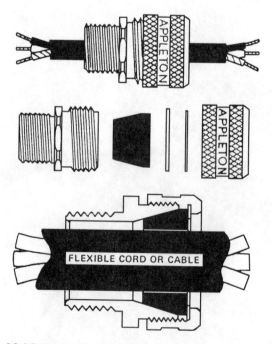

Fig. 10-10 *Watertight cable connectors. These are special application connectors for use in special areas in residential buildings. They are used mostly in commercial and industrial applications.*

A hub is an enlargement on the end of a piece of pipe or conduit. This means another pipe or conduit of the same size may be inserted and sealed with a satisfactory joint. In this case, a hub means that the end of the connector is such as to accept a piece of conduit of the size specified on the connector. Installation of watertight connectors is easy (Fig. 10-11). These connectors have standard, tapered pipe threads for installation directly into a threaded cast fitting. A locknut may be added if the cable is to be used on an enclosure that is not threaded. This may occur when it is used on a outlet box or a steel junction box.

BX CABLE AND THE CODE

BX cable has an official designation of AC or armored cable. This type of cable is now covered by its own article in the Code. It has been drawn out and separated from the nonmetallic cables or Romex. This type of cable, according to Underwriters Laboratories, has a copper or aluminum bonding strip in contact with the armor. The covering is made of steel. There are insulated conductors inside the cable for current-carrying purposes.

Armored cable can be obtained in two-, three-, or four-conductor sizes varying from No. 14 to No. 1. All these will conform to the UL specifications. It is for use at temperatures of 60°C (140°F) or 75°C (167°F), depending upon its insulation over the conductor. This type is used for voltages up to 600. Other metal-covered cables are not covered by this article of the Code. You should read the cable description on the box when purchasing.

One of the requirements of the Code for BX is the insertion of an antishorting bushing, sometimes called a *red head.*

Types of Armored Cable

Certain letters are used to designate the type of cable with UL data on the conductors:

The letters ACT are used to indicate armored cable that has conductors with thermoplastic insulation. This type of insulation is usually designated with a T. Therefore, ACT designates armored cable with thermoplastic insulation.

AC indicates armored cable with conductors that are covered with rubber insulation of the grade specified by the Code.

ACH indicates armored cable that has rubber insulation that will withstand heat within the 75°C (167°F) range.

ACHH indicates cable with a temperature range of 90°C (194°F).

ACU indicates armored cable with rubber insulation, but of the latex grade.

Whenever L appears as a suffix, it indicates that lead has been used as a covering over the conductors.

Uses of Armored Cable

Armored cable is used for branch circuits, feeders, and light circuits. You have to be careful with the wiring of BX because it is possible for induction heating to cause problems. Induction heating takes place when the current feeding a particular device comes through a conductor in one metal-covered wire and returns to the source in another metal-covered wire inside a parallel arrangement of the two metal covers that also service as conductors (Fig. 10-12).

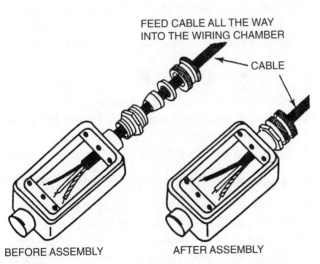

Fig. 10-11 *Watertight connectors in a unilet. The unilet is usually associated with conduit installations. In some home installations this can be a very useful arrangement.*

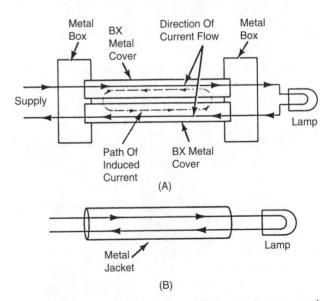

Fig. 10-12 *Induction can be caused if the wire is not properly pulled through a conduit or BX.*

Induction Induction takes place whenever a magnetic field cuts across a conductor. In this case, the magnetic field is established by the current-carrying wires. If the current flows as shown in Fig. 10-12A, then the magnetic field creates a path for an induced current in the metal BX covering and the junction boxes. This can be sufficient, at 30 amperes and over, to reduce severely the voltage available at the lamp. It may also generate heat that can cause a fire in anything contacting the heated metal covering or armored case.

Figure 10-12B shows the normal condition of wiring in conduit or BX. Both conductors are housed in the same metal jacket. The current flowing to the lamp and the current flowing away from the lamp are in different directions. They have a tendency to cancel the magnetic field generated by the current-carrying wires.

Type AC is used for signal and control circuits. Some installations of speakers in remote locations are common. It is used in new construction and in remodeling. However, it is more expensive initially than Romex. This factor alone limits its use in home construction and wiring.

Type AC or even ACL cable cannot be used where it is buried directly in the earth.

Armored cable must be secured by approved staples, straps, or fittings. The straps or staples should not be spaced more than 24 to 30 inches apart. If the wire is concealed, you can support it every 4½ feet with staples or an appropriate fastening device. It should be secured within 1 foot of each outlet box or fitting.

Armored cable should be supported every 2 feet. This applies in particular to lamp fixtures or any enclosure or equipment. However, Greenfield flexible tubing and liquidtight tubing can be left for as much as 6 feet without support. This does not apply to BX. BX must be supported every 2 feet.

Whenever you use another wiring method and it connects to a BX cable, the box must be a C conduit body type (Fig. 10-13).

INSTALLING ELECTRIC SERVICE WITH CONDUIT

Conduit is not too hard to install. It requires a conduit bender, hacksaw, screwdriver, hammer, and pliers. In some areas, the Code requires conduit for home use. This varies from place to place. Thus, it is best to check with your local power company to make sure. It is the most expensive type of wiring. If the building has more than three floors above grade level, Romex or nonmetallic cable cannot be used. Other types must be used—either BX or conduit.

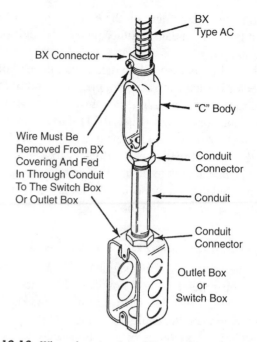

Fig. 10-13 When changing from BX to conduit or some other type of wiring method, you must use a C body for the transfer.

Wires from the utility company's pole to your building are called a service drop. This is usually furnished by the company. These wires must be high enough to provide proper clearance above grade. They must not come within 3 feet of doors, windows, fire escapes, or any opening. The structure to which the service drop is fastened must be sturdy and able to withstand the pull of ice, wind, and other weather conditions.

Figure 10-14 shows some of the clearances for a service drop. These are in accordance with the National Electrical Code. Figure 10-15 shows how the required clearance is figured. In some cases, there are exceptions. Figure 10-16 shows one of the exceptions. This one is brought about by the length of the overhang.

Figure 10-17 indicates some of the details that must be considered when you install a service drop.

Installation Procedures

Attach the service head to the building as suggested by the Code and your local regulations. Connect the conduit as shown in Fig. 10-18. Use a metal strap every 4 feet to fasten the conduit to the building. Connect the meter with conduit connectors. Use an entrance ell to turn the conduit into the house. The ell has two threaded openings corresponding to conduit size. Use an adapter to fasten the conduit into the threaded opening at the top of the ell. Into the lower opening fasten a piece of conduit to run through the side of the house. Use a connector to attach the conduit to the electric service panel (Fig. 10-18).

Fittings The following fittings are used.

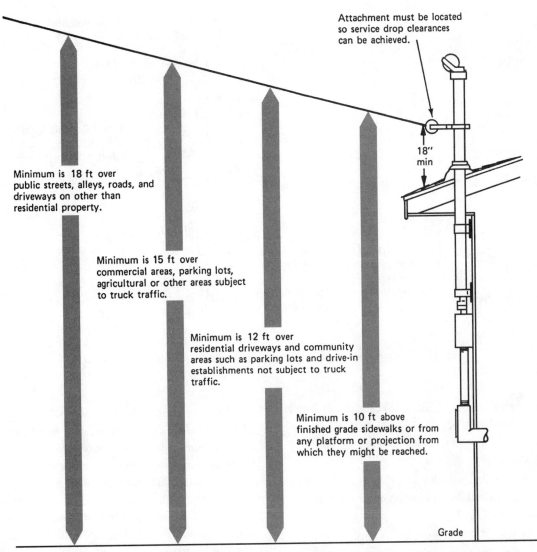

Attachment must be located so service drop clearances can be achieved.

18" min

Minimum is 18 ft over public streets, alleys, roads, and driveways on other than residential property.

Minimum is 15 ft over commercial areas, parking lots, agricultural or other areas subject to truck traffic.

Minimum is 12 ft over residential driveways and community areas such as parking lots and drive-in establishments not subject to truck traffic.

Minimum is 10 ft above finished grade sidewalks or from any platform or projection from which they might be reached.

Grade

Fig. 10-14 *Various clearances for service drop.*

Entrance elbows These are used customarily where the conduit enters a building. The size of the conduit and fittings depends on the size of the entrance wires.

Entrance elbows must be raintight. Before you pull the wires through the conduit, remove the cover on the ell. (*Ell* is short for *elbow.*) Use a fish wire to pull the wires into place. After connection of the wires, replace the cover. Entrance ells come in almost any shape and size to fit almost any installation (Fig. 10-19).

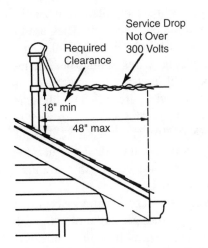

Required Clearance

Service Drop Not Over 300 Volts

18" min

48" max

Fig. 10-15 *Mounting the service head on the house.*

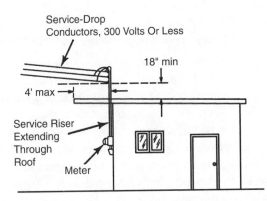

Service-Drop Conductors, 300 Volts Or Less

18" min

4' max

Service Riser Extending Through Roof

Meter

Fig. 10-16 *One exception to the Code requirements for the location of a service drop.*

Installing Electric Service with Conduit   167

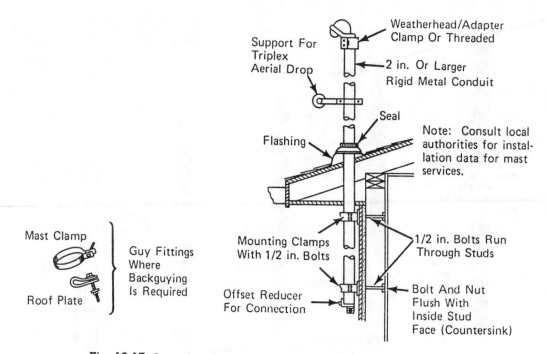

Support For Triplex Aerial Drop

Weatherhead/Adapter Clamp Or Threaded

2 in. Or Larger Rigid Metal Conduit

Seal

Flashing

Note: Consult local authorities for installation data for mast services.

Mounting Clamps With 1/2 in. Bolts

1/2 in. Bolts Run Through Studs

Offset Reducer For Connection

Bolt And Nut Flush With Inside Stud Face (Countersink)

Mast Clamp

Roof Plate

Guy Fittings Where Backguying Is Required

Fig. 10-17 *Proper installation of a service drop and mounting the weatherhead.*

Ninety-degree female elbows These are designed for use with conduit-to-conduit, right-angle installations. Elbows are generally used instead of bending the conduit itself (Fig. 10-20).

Ninety-degree bushed elbows These are used to make right-angle installations, either conduit-to-conduit, or conduit-to-box (Fig. 10-21).

Pulling elbows These provide a simple yet effective means of pulling wires through raceways. They are used in cases where it is necessary to run the conduit around 90° corners (Figs. 10-22 and 10-23). The pulling elbows illustrated have removable covers which, when removed, allow you to pull the wires freely all the way through one side of the fitting, then through the other side. The cover is then replaced.

Forty-five-degree female conduit elbows These are designed for use when a conduit-to-conduit, 45° angle installation is required. These elbows are generally used instead of bending the conduit (Fig. 10-24).

Steel-cupped reducing washers These are used with outlet boxes. They reduce the diameter of a knockout to a smaller opening. They are made of cadmium-plated steel and are used in pairs. One is placed inside the outlet box, the other outside. Conduit is placed through them with a locknut on either side (Fig. 10-25). Tighten the locknuts to complete the installation. A ledge on the edge of the washer prevents it from slipping after it has been installed.

Entrance caps for use with threaded rigid conduit These are also known as *service heads*, *entrance fittings*, *entrance heads*, or *weatherheads*. They are used to bring electrical power into a building. Current-carrying wires are brought from the outdoor lines to the entrance cap. The entrance cap is anchored on the building. Wires are then pulled from the service switch inside the building and through the conduit. The conduit passes from the inside wall to the outside wall of the building. The wires are pulled into an entrance elbow mounted on the exterior wall of the building (Fig. 10-26).

Conduit comes out of the building into an ell. Then conduit is connected from the ell (located at the bottom of the building or foundation) to the service entrance cap. Wires are pulled from the ell, through the conduit into the entrance cap. They are then threaded through the wire holes in the insulating block of the service cap. The wires should be extended about 18 inches (455 millimeters) and cut off. At this point, the interior circuit wires are spliced to the exterior wires.

Several types of service caps are available. All have removable covers, in which a detachable insulating block keeps the individual circuit wires separated. Some have snap-on covers that eliminate attaching screws on the cover.

In some cases, service entrance cable is used instead of conduit. For these installations, the cable entrance caps have a clamp neck instead of threads or setscrews.

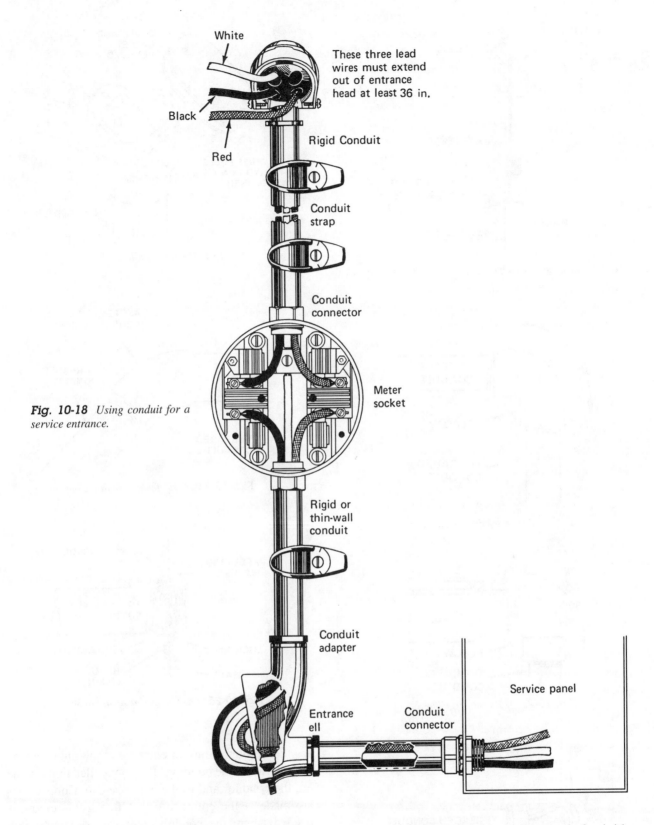

White

These three lead wires must extend out of entrance head at least 36 in.

Black

Red

Rigid Conduit

Conduit strap

Conduit connector

Meter socket

Fig. 10-18 *Using conduit for a service entrance.*

Rigid or thin-wall conduit

Conduit adapter

Service panel

Entrance ell

Conduit connector

The cap is fastened to the building instead of being supported by the conduit. The cable is secured to the building with clamps or straps (Fig. 10-27).

Thin-wall entrance caps and elbows These are like the threaded rigid conduit caps and elbows. How-

ever, they are made for thin-wall instead of a rigid conduit. They are used, as previously stated, to bring power into a building. The cables are brought from the overhead supply outdoors to the entrance cap anchored on the building. Wires from the inside of the building

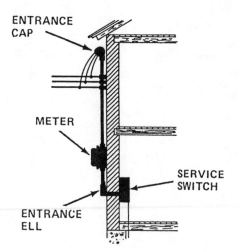

Fig. 10-19 *Entrance cap, meter, entrance ell, and service switch.*

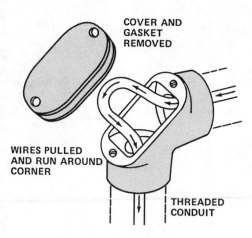

Fig. 10-23 *A pulling elbow.*

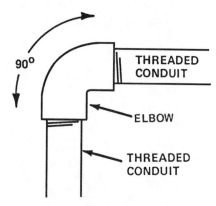

Fig. 10-20 *A 90° female elbow.*

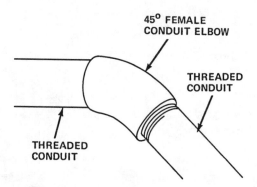

Fig. 10-24 *A 45° female conduit elbow.*

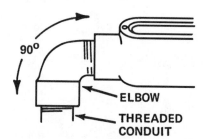

Fig. 10-21 *A 90° bushed elbow.*

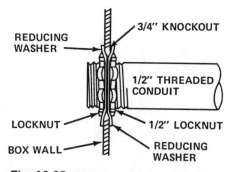

Fig. 10-25 *Steel cupped reducing washers.*

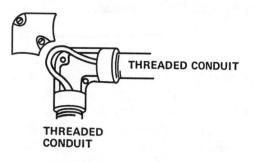

Fig. 10-22 *A 90° pulling elbow.*

are then brought into an entrance elbow mounted on the exterior of the building. They are pulled through the ell, the conduit, and to the entrance cap. Connections are made at the entrance cap. Often a 90° elbow is needed where the conduit enters the service switch (Figs. 10-28 and 10-29).

Another type of entrance ell is shown in Fig. 10-30. It mounts to EMT with compression rings and nuts on either end. A removable cover facilitates the pulling of wires from one end and feeding them through the other.

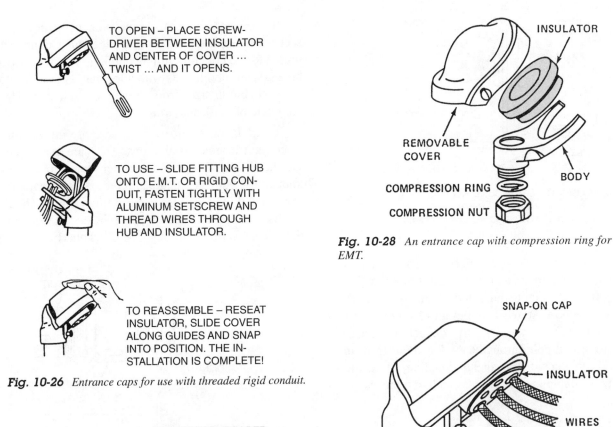

TO OPEN – PLACE SCREW-
DRIVER BETWEEN INSULATOR
AND CENTER OF COVER ...
TWIST ... AND IT OPENS.

TO USE – SLIDE FITTING HUB
ONTO E.M.T. OR RIGID CON-
DUIT, FASTEN TIGHTLY WITH
ALUMINUM SETSCREW AND
THREAD WIRES THROUGH
HUB AND INSULATOR.

TO REASSEMBLE – RESEAT
INSULATOR, SLIDE COVER
ALONG GUIDES AND SNAP
INTO POSITION. THE IN-
STALLATION IS COMPLETE!

Fig. 10-26 *Entrance caps for use with threaded rigid conduit.*

INSULATOR

REMOVABLE
COVER

COMPRESSION RING

COMPRESSION NUT

BODY

Fig. 10-28 *An entrance cap with compression ring for EMT.*

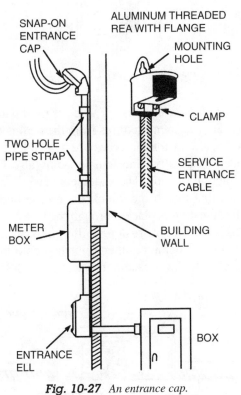

SNAP-ON
ENTRANCE
CAP

ALUMINUM THREADED
REA WITH FLANGE

MOUNTING
HOLE

CLAMP

TWO HOLE
PIPE STRAP

SERVICE
ENTRANCE
CABLE

METER
BOX

BUILDING
WALL

BOX

ENTRANCE
ELL

Fig. 10-27 *An entrance cap.*

SNAP-ON CAP

INSULATOR

WIRES

SETSCREW

CONDUIT

Fig. 10-29 *An entrance cap with setscrew mounting and snap-on cap.*

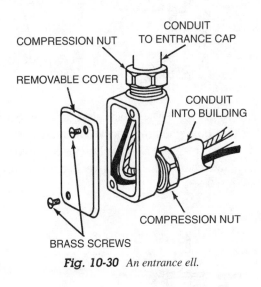

COMPRESSION NUT

CONDUIT
TO ENTRANCE CAP

REMOVABLE COVER

CONDUIT
INTO BUILDING

COMPRESSION NUT

BRASS SCREWS

Fig. 10-30 *An entrance ell.*

Another 90° elbow is shown in Fig. 10-31. It is available in a long or short size. You use it to connect a run of thin-wall conduit to a metal enclosure with a knockout at a 90° angle to the conduit fan. The EMT

is held in the female end with a compression ring and nut. The male end is secured to the entrance with a tiger-grip locknut. The edges of all these fittings are chamfered to prevent possible damage to wire insulation and cable sheath.

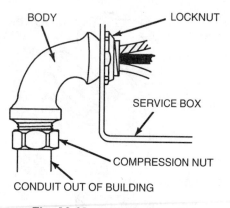

Fig. 10-31 *A 90° elbow installed.*

Straight nipples or chase nipples These are used with conduit couplings to attach conduit to a box (Fig. 10-32). Place the nipple through the knockout from the inside of the box. Attach a coupling to the nipple, and the conduit to the coupling. To connect two boxes, place a chase nipple through each knockout from inside and connect with the conduit coupling. Chase nipples are available in sizes from ³/₈ inch to 5 inches.

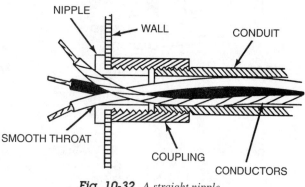

Fig. 10-32 *A straight nipple.*

Offset nipples These are designed to connect conduit to a box when the knockouts and conduit run do not line up. They are also used to connect two boxes side by side when their respective knockouts do not align (Fig. 10-33).

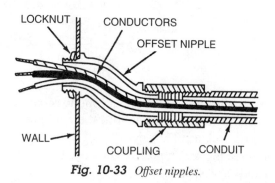

Fig. 10-33 *Offset nipples.*

To connect conduit to a box using an offset nipple, secure the nipple to the box with a locknut. Connect the nipple to the conduit with a coupling. To connect two boxes, secure both ends with a locknut. Offset nipples are available in sizes from ¹/₂ inch to 1¹/₄ inches. They are made of malleable iron.

Conduit male enlargers These are vital fittings in conduit-to-box installations. They are used where the diameters of the conduit differ from those of the knockouts. They are made, in three sizes, of malleable iron. Plated over the iron is a cadmium finish. The fitting is installed by threading over the end of rigid conduit, then inserting it through the knockout (Fig. 10-34).

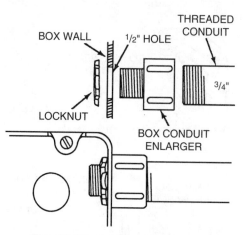

Fig. 10-34 *Conduit male enlargers.*

No-bolt fixture stems These are used with outlet boxes. They suspend fixtures from the ceiling without bolts. Male and female types are available. Both are threaded externally. The female stems also have ³/₈-inch internal threads. The shoulders have ¹/₈-inch protrusions that match the knockouts on the outlet box. The fixture stem is placed through the base of the outlet box with the protrusions falling into place in the knockouts. A locknut is then tightened onto the stem. This holds the fixture stem in place (Fig. 10-35). Both

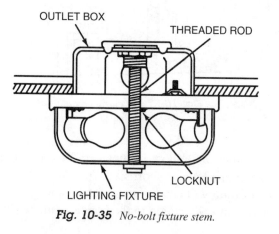

Fig. 10-35 *No-bolt fixture stem.*

types are made with a cadmium finish to prevent rust and corrosion.

Tiger-grip locknuts These attach a piece of conduit or a connector into a knockout opening in a metal box. One inside locknut is used in connecting conduit to a box. Another locknut is used on the outside of the box. The pressure exerted by the two locknuts locks the conduit firmly in place through the wall (Fig. 10-36).

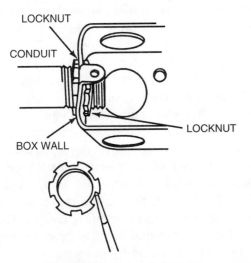

Fig. 10-36 *Tiger-grip locknuts.*

In tiger-grip locknuts there are nonslip notches around the perimeter of the locknut. These notches help prevent wrenches and screwdrivers from slipping. They also provide a perfect ground when tightened. You can be sure of the connection since their tilted edges bite into the metal wall of the box. The connection will not vibrate loose from excessive movement. Tiger-grip locknuts have been used where local codes demand an "approved" locknut.

Snap-in blanks for knockouts These are used to seal off knockouts made in error, or in rework in metal boxes (Fig. 10-37).

Rigid conduit hubs These consist of two parts: the body and the hex-head wedge adapter. The threaded

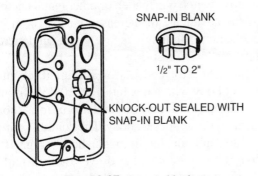

Fig. 10-37 *Snap-in blank.*

shank of the hex-head wedge adapter is placed through the knockout from the inside of the box. The hub body is screwed onto the wedge adapter until the fitting is reasonably tight with the box. Conduit is then installed to the hub and the fittings tightened securely (Fig. 10-38). The hex-head design permits tightening from either outside or inside the enclosure (Fig. 10-39).

Often, when making an installation using threaded rigid conduit, junction boxes and device housings are used which do not provide suitable means for fastening the conduit to the enclosure. Knockouts, found in most enclosures, provide a location where rigid conduit hubs can be fastened. If there are no knockouts in the enclosure, a hole must be drilled.

The self-locking, hex-head wedge adapter and the hub body exert a continuous, uniform 360° pressure on the inside and outside surfaces of the box wall. This eliminates the need for locknuts. All hubs have a built-in, recessed neoprene gasket and a flame-resistant insulated throat. The latter eliminates the need for an end bushing. It protects the wire insulation and cable sheath from damage due to vibration. The insulated throat leaves more wiring room within the enclosure.

Connectors are made with a zinc finish over steel. Larger sizes are of malleable iron with a cadmium finish.

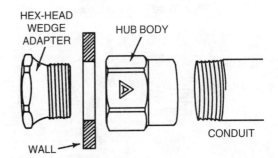

Fig. 10-38 *Hex-head wedge adapter and hub body for rigid conduit.*

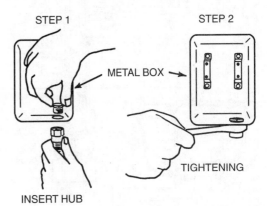

Fig. 10-39 *Attaching a hub to the box and adapter.*

Thin-Wall Conduit

Thin-wall conduit is metal tubing. It is called *electrical metallic tubing* (EMT) by those who use it. This type of wiring is used with metal boxes only. There are connectors made for this type of wiring. They will be described and shown being attached in the following paragraphs.

Empty EMT should be mounted in place and connected to boxes; then the wire is pulled through it (Figs. 10-40 and 10-41).

Figure 10-42 shows one method of using conduit to wire a house. Note how the studs are notched to allow the conduit to be installed. Lengths shorter than 10 feet are cut with a hacksaw or tubing cutter. The ends are reamed so the cut ends are tapered and no jagged edges are left to cut through the insulation on the conductors.

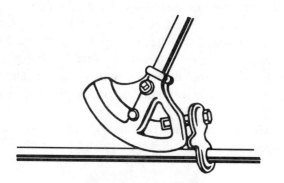

Fig. 10-40 *Conduit bender.*

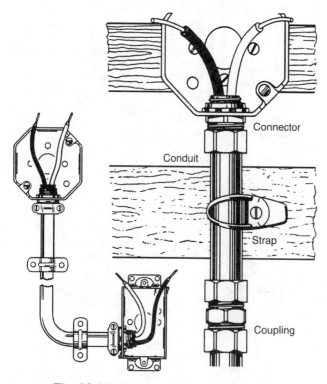

Fig. 10-41 *Conduit with wires pulled through.*

Conduit should be supported with a strap every 6 feet on exposed runs. In some instances it would be better to support it at shorter intervals.

When you are connecting conduit to boxes, fit the threadless end of the conductor over the conduit, and insert the connector through the box knockout. Then tighten the locknut.

After conduit and boxes are installed, pull the wires through the conduit into the boxes. Allow 8 inches of insulated wire at each end of the box for connections. Use a white wire for neutral and a black or red wire for the "hot" side.

In exposed work, conduit may be mounted on studs or rafters without additional protection. In concealed work, conduit must be supported.

Thin-wall conduit compression connectors and couplings are used with a special EMT (thin-wall) conduit. Thin-wall conduit connectors are employed in fastening (connecting) EMT to outlet boxes, switch boxes, panels, and other metal enclosures (Fig. 10-43). Thin-wall conduit *couplings* are electrical fittings used to attach (couple) the length of one conduit to another. They are used to extend the overall length of a piece of conduit. Conduit usually comes in 10-foot lengths.

Throat openings of the thin-wall connectors and couplings are chamfered to eliminate burrs. That means that ends have been slightly reamed out to eliminate small metal bits that can damage wire being pulled through the conduit. All sizes of these connectors and couplings have hexagonal nuts, and bodies that must be held securely. A wrench is used to hold the nuts when tightening. The body of the fitting has a conduit stop that allows the EMT to enter evenly for uniform strength.

The ends of the connector, or coupling, house a compression ring. When the nut is tightened, the inward motion of the nut forces the open compression ring into a closed position around the conduit. This action locks the fitting onto the conduit.

Two-piece thin-wall conduit connectors   These are used where thin-wall conduit is to be connected to an outlet box, switch box, panel, or other metal enclosure. It has the advantage of increased wiring room inside the enclosure. Installation is fast with a single wrench. This means lower cost (Fig. 10-44).

This connector consists of a knurled, chamfered split-steel body with a hex nut (Fig. 10-45). To install, insert the body through the outlet box to a shoulder stop rim on the connector body. The conduit is then inserted through the nut into the body (Fig. 10-46). Tighten with a wrench placed over the hex nut. This draws the knurled chamfer of the body against the

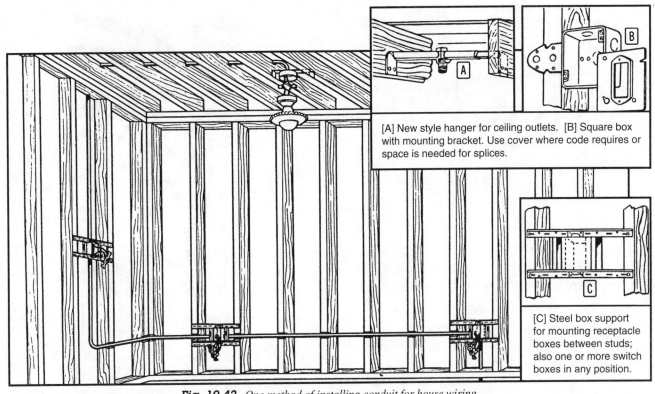

[A] New style hanger for ceiling outlets. [B] Square box with mounting bracket. Use cover where code requires or space is needed for splices.

[C] Steel box support for mounting receptacle boxes between studs; also one or more switch boxes in any position.

Fig. 10-42 One method of installing conduit for house wiring.

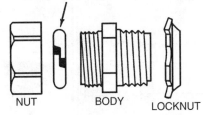

COMPRESSION RING

NUT   BODY   LOCKNUT

Fig. 10-43 Compression connector.

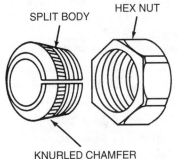

SPLIT BODY   HEX NUT

KNURLED CHAMFER
SHOULDER STOP

Fig. 10-45 Two-piece conduit connector.

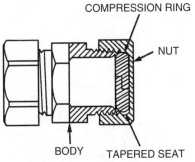

COMPRESSION RING

NUT

BODY   TAPERED SEAT

Fig. 10-44 Pressure-cast coupling for EMT.

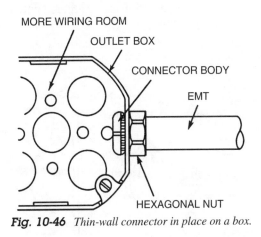

MORE WIRING ROOM

OUTLET BOX

CONNECTOR BODY

EMT

HEXAGONAL NUT

Fig. 10-46 Thin-wall connector in place on a box.

knockout hole of the box. At the same time, it compresses around the conduit. Thus, slippage is prevented, and there is no need for a locknut.

Two-piece, thin-wall conduit connectors may be used to install conduit between two stationary enclosures. If used here, simply cut the conduit to the correct length. Slip the hex nuts on both ends. Insert the conduit

through the knockout in one box far enough that it may be backed up through the knockout in the second box. (Backing up is shown in Fig. 10-47.) Place the

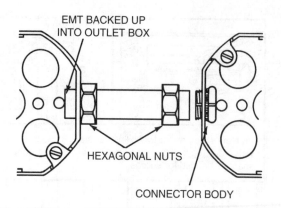

Fig. 10-47 *Two-piece connectors used between two boxes.*

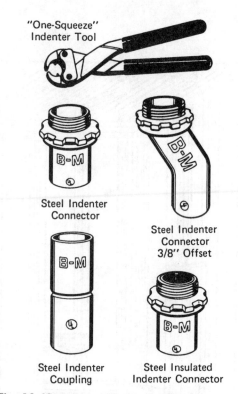

Fig. 10-49 *Indenter connectors and an indenter tool.*

connector bodies over the ends of the conduit from the inside of the box. Tighten with a single wrench. These connectors are concrete-tight and UL-approved.

Indenter-type thin-wall conduit connectors and couplings These are used for attachment of thin-wall conduit to metal enclosures. They are designed for joining two sections of thin-wall. They provide a permanent, rigid connection or coupling. Install them with an indenter tool (Figs. 10-48, 10-49, and 10-50). All indenter fittings feature chamfered edges to prevent damage to cable sheath and wire insulation. Some have insulated throats.

Install by inserting the EMT into the fitting until it rests against the conduit stop. This allows the conduit to enter only to the halfway point. Here it is held in check. This provides the strongest possible connection. With the EMT inserted, the indenter tool jaws are placed around the fitting and squeezed tightly (Fig. 10-51). The prongs make deep indentations in the fitting and conduit. The tool is then rotated 90°, and another set of indentations is made.

Rotation and indentation are continued until a total of four have been made. The process is repeated for completion of coupling installation (Fig. 10-52).

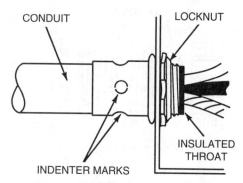

Fig. 10-50 *Indenter connector installed.*

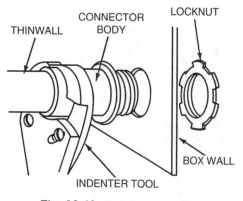

Fig. 10-48 *An indenter coupling.*

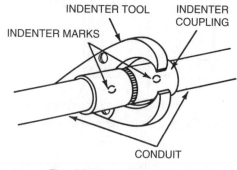

Fig. 10-51 *Indenter coupling.*

Indenter-type connectors have a tiger-grip locknut. The locknut is tightened on the inside of the enclosure for a slipproof bond (Fig. 10-50). They are, of course, plated with cadmium to prevent rust or corrosion.

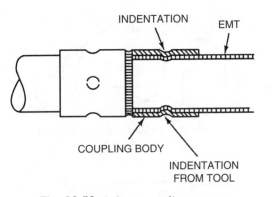

Fig. 10-52 *Indenter coupling cutaway.*

Setscrew thin-wall conduit connectors and couplings
These are used where EMT is to be fastened (connected) to a switch, outlet, or panel box. Setscrew couplings are applied where one run of thin-wall conduit is to be attached (coupled) to another. These connectors and couplings are designed for use on straight runs of conduit, that is, where two ends of the thin-wall meet and line up properly. They may also be used where the two ends of the conduit line up with the knockouts of a box. Figures 10-53 and 10-54 show a setscrew connector.

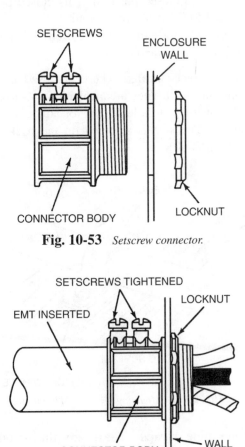

Fig. 10-53 *Setscrew connector.*

Fig. 10-54 *Installed setscrew connector.*

To install, simply loosen the setscrews on the fitting. Then insert the thin-wall conduit until it falls against the built-in stop collar inside the fitting. This stop collar ensures even holding pressure for EMT. Also, it provides a smooth surface over which wires may be pulled easily without damage to the insulation. All edges are chamfered for further protection.

After inserting the conduit, tighten the setscrews against the outer wall of the conduit. Deep-slotted, staked setscrews thread firmly into the embossed surface of the body of the fitting. Repeat the procedure to install the conduit in the other side of the coupling. To complete the connector installation, insert the conduit and the connector in the box knockout. Tighten the locknut to the interior of the box to prevent slippage (Fig. 10-55).

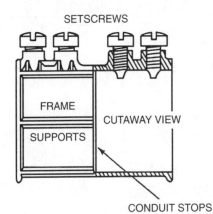

Fig. 10-55 *Cutaway view of a coupling, setscrew type.*

The combination flexible-steel conduit to EMT coupling is manufactured in 1/2-, 3/4-, and 1-inch sizes. One end has a split body with a tightening screw for flexible-steel conduit to a thin-wall fitting. Place the end with the split body over the flexible conduit. Tighten the screw (Fig. 10-56). Then insert the thin-wall conduit in the compression ring end and tighten the nut for a rigid bond.

A *conduit stop* inside the body of the fitting ensures an even strength of bond. All edges are chamfered to prevent damage to cable sheath and wire insulation. Heavy-gage steel fittings are precision-machined. They have a cadmium finish to resist rust and corrosion. At the open end of the clamp is a lip that locks conduit into place. An oval screw hole allows adjustment after installation. Sizes are stamped on the back of the clamp for easy identification (Fig. 10-57).

Two-hole pipe clamps Two-hole pipe clamps, or straps, are flush-mounted supports for fastening conduit to a wall or ceiling. They are light in weight, yet durable and versatile. On both sides of the arch of the two-hole pipe strap are protruding punch marks that

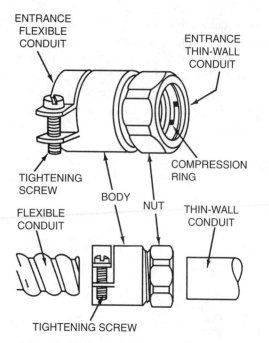

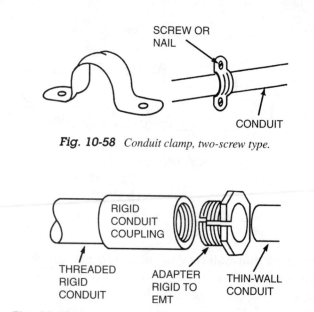

Fig. 10-58 *Conduit clamp, two-screw type.*

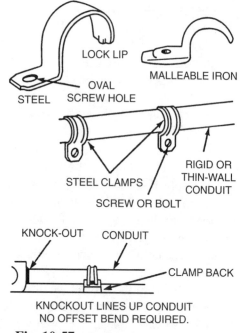

Fig. 10-56 *Thin-wall to flexible-steel fitting.*

Fig. 10-59 *Installation of threaded rigid conduit to thin-wall conduit adapters.*

The adapter is a compression-type fitting with external threads: It has a split body. Thin-wall conduit is placed inside the adapter, and the assembly is tightened into the threaded fitting. As you tighten the adapter, it is forced into a closed position. This closed position around the EMT forms a rigid bond.

Conduit clamps and straps are used to fasten conduit to walls, floors, and ceilings.

One-screw steel clamps These are used in flush-mounting thin-wall. They are available in $1/2$- to 2-inch sizes. An oval screw hole allows for adjustment after installation. The size is stamped on the back of the clamp (Fig. 10-60).

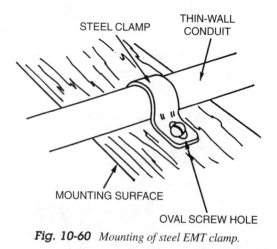

Fig. 10-60 *Mounting of steel EMT clamp.*

Fig. 10-57 *Conduit clamp, one-screw type.*

serve to lock the conduit or cable in place. Identification is easy. The size is marked on the back of each strap (Fig. 10-58).

Adapters, clamps, and straps for thin-wall conduit These are made for a number of jobs. The adapters are used for converted threaded, rigid-conduit fittings for use with thin-wall conduit (Fig. 10-59).

Two-hole pipe straps These are flush-mounting supports for thin-wall. They are inexpensive and are

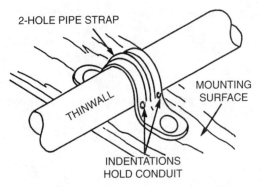

Fig. 10-61 *A two-hole pipe strap installed.*

available in sizes up to 1½ inches. Protruding marks on both sides (raised above the surface) of the strap serve to lock the conduit in place. The size is stamped on the back of the strap (Fig. 10-61).

Note that there is a distinct difference between clamp diameters for rigid conduit and those for thinwall conduit.

Bushings

End bushings are used where threaded rigid conduit enters a metal enclosure. These enclosures may be boxes for controls, panel boards, or fuse boxes. A method to protect the wire from abrasion must be provided. This is necessary because the end of the conduit is very rough and is therefore capable of cutting the conductors. This cutting of the conductors causes a dead short or other damage. For this purpose, end bushings are designed to thread onto the end of the conduit so that the teeth will bear against the outlet box in which the conduit is fastened.

The end of the conduit is then slipped through the knockout (Fig. 10-62). There is a raised, rounded surface in the bushing over which the wire slides. The wire does not touch the conduit at this end (Fig. 10-63). After the bushing is installed, tighten the locknut on the outside of the box. The teeth of the locknut dig into the metal box, making a continuous ground.

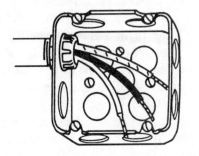

Fig. 10-62 *Locknut and bushing properly installed.*

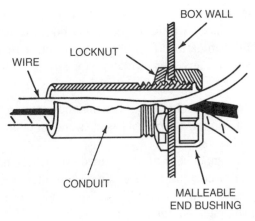

Fig. 10-63 *Thermoplastic bushings.*

There are several types of end bushings available. For example, malleable-iron bushings secure the conduit in the same manner as a locknut. Plastic bushings protect the end of the conduit, but require the addition of locknuts for holding.

CONDUITS AND THE CODE

EMT is a general-purpose raceway. It is similar to rigid conduit and intermediate metal conduit (IMC). It is between rigid conduit and IMC in wall thickness. It does use the same threading method and standard fittings as rigid metal conduit. Generally speaking, it also fits the bill for rigid conduit wherever it is called for. The only difference is that IMC is slightly lighter, since its wall thickness is less than that of rigid conduit. It requires about 25% less steel than rigid conduit. It performs about as well as rigid in most places. It is better, of course, than rigid aluminum and EMT in most locations. It is less expensive than rigid, but more expensive than EMT. Its outside diameter is the same as that of rigid conduit. This larger inside diameter makes it easier to pull wires through. However, the NEC does not allow for more wires to be included because of its greater inside diameter.

Comparison of Rigid Conduit and IMC

The ¾-inch trade sizes of both rigid conduit and IMC are shown in Table 10-1.

Table 10-1 *Comparison of Rigid Conduit and Intermediate Metal Conduit*

Conduit	Outside Diameter	Inside Diameter	Wall Thickness
Rigid steel	1.050 inches	0.824 inch	0.113 inch
IMC	1.050 inches	0.908 inch	0.071 inch

Clamping of IMC   A clamp or other type of support must be placed within 3 feet of any box, cabinet, or conduit termination. There shall be not more than 10 feet between any pair of clamps for any IMC at any angle.

Number of conductors   The number of wires allowed in rigid, IMC, or EMT is the same for conduit ½ inch and larger.

IMC cannot be used when buried in cinders. Cinders have sulfur content. If there is moisture, then the cinders present a sulfuric acid. This acid corrodes the metal. The IMC must be buried at least 18 inches below the cinder fill. However, it can be buried in the cinders if encased in 2 inches of concrete.

A run of conduit cannot have more than 360° of bends from the point of outlet to outlet. This includes the bends at the outlets. For example, if you have a 45° angle at the outlet box and another 45° so the conduit fits against a flat surface, this counts as 90°. If the same bends are used at the other box, then another 90° is used. That leaves only 180° of bends for the entire length of the run from box to box or panel to box.

Aluminum conduit definitely requires additional corrosion proofing if it is buried in concrete. Galvanized steel conduit, rigid conduit, and galvanized IMC installed in concrete do not require supplementary corrosion treatment.

Rigid nonmetallic conduit   Materials used for this type of conduit are fiber, asbestos, cement, soapstone, rigid polyvinyl chloride, and high-density polyethylene for underground use. Rigid polyvinyl chloride is used aboveground.

Plastic conduit is a better name for rigid nonmetallic conduit. It is permitted to be used where the voltage does not exceed 600 volts. It can be used in walls, floors, and ceilings. This type of conduit lends itself for use in cinder fill situations. However, if the voltage is over 600 volts, it must be encased in 2 inches of concrete.

Plastic conduit must be used where the temperature does not exceed its rating. It should not be used where there is the possibility of physical damage. It should not be used to support fixtures or other equipment. Support of the conduit must be as shown in Table 10-2. In addition, note that there shall be a support within 4 feet of each box, cabinet, or other conduit termination.

Polyvinyl chloride conduit and high-density polyethylene conduit are made and shipped in 10-foot lengths. Conduit bodies must have a cross-sectional area at least twice that of the largest conduit to which they are connected.

The number of conductors in conduit is limited by the National Electrical Code. For new work or rewiring, the total cross-sectional area of the conductors must not be over 40% of the internal cross-sectional area of the conduit. For three or more conductors, the sum of their areas must not exceed 40% of the conduit area. For details on the fill of conduit, check the latest issue of the National Electrical Code.

Aboveground Use

Polyvinyl chloride (PVC) conduit is the only rigid *nonmetallic* conduit that may be used aboveground. There are two schedules of rigid nonmetallic conduit—Schedule 80 and Schedule 40. Schedule 80 has an inside diameter that is less than that of Schedule 40. For conductor fill to 40% of the cross-sectional area, the wire fill capacity is marked on the conduit surface.

This plastic PVC conduit can be used aboveground for over 600 volts if it is encased in at least 2 inches of concrete. It can be buried if it meets the depth requirements of the Code. This type of conduit can be obtained with a diameter of ½ inch to 6 inches. One coupling is furnished with the 10-foot length of conduit. This type of conduit can be used where there are corrosive agents, such as in industry. It is also usable where there are vapors, caustic mist, pickling acids, and plating baths. In the home you would probably use it to mount spotlights outdoors on poles or along the top edges of the house. This type of conduit is cemented together at junctions or couplings with cements made for the specific type of plastic.

There are factory-made elbows and fittings for the PVC and other types of plastic or nonmetallic conduit (Fig. 10-64).

Table 10-2 *Support for Rigid Nonmetallic Conduit*

Conduit Size (Inches	Maximum Spacing between Supports (Feet)	
	Conductors Rated 60°C (140°F) and Below	Conductors Rated More Than 60°C (140°F)
½ to ¾	4	2
1 to 2		5 2½
2½ to 3	6	3
3½ to 5	7	3½
6		8 4

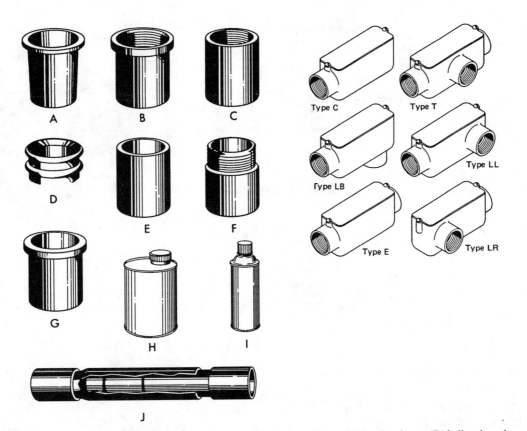

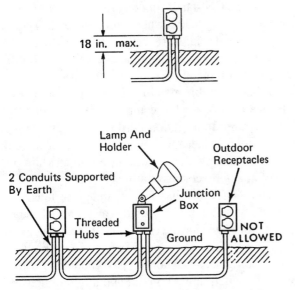

Fig. 10-64 *Factory-made fittings for PVC conduit. (A) Bell ends and caps, (B) reducer, (C) female adapter, (D) bell ends and caps, (E) coupling, (F) terminal adapter, (G) adapter, (H) PVC cement (quart size), (I) PVC cement (pint size), (J) expansion joint. The access fittings on the right are identified by type.*

Hotbox benders are available for bending the plastic conduit. It simply heats the plastic so it can be bent to the desired angle.

Take a look at Fig. 10-65. This is the standard mounting for a box if it is supported by two threaded

Fig. 10-65 *Mounting boxes and lamps on conduit.*

conduits stubbed up out of concrete or the ground. However, it cannot be used for nonmetallic conduit.

These limitations must be kept in mind for residential wiring. Sometimes it is desirable to have a patio plug or flowerbed plug available for operating power equipment. The maximum height of 18 inches helps the pipe or conduit withstand blows that may be accidentally applied.

FLEXIBLE CONDUIT

This type of conduit is known as Greenfield or flex (Fig. 10-66). It comes in coils no longer than 250 feet. However, it can be bought in coils of 50 and 100 feet. The $5/16$-inch size as well as the $3/8$-inch size comes in 250-foot coils. The $1/2$-inch size is available in 100-foot coils, and the other three sizes ($3/4$-, 1-, and $1 1/4$-inch) come in rolls of no more than 50 feet.

It may be used in runs no longer than 6 feet with no requirement for a separate ground wire. Flex can be obtained in either aluminum or steel. It has been judged to be suitable for a ground only up to 6 feet in length. When you use the aluminum flex, you must be careful not to tighten the connectors with setscrews too tightly.

FLEXIBLE STEEL CONDUIT

U.L. & C.S.A LABELED FLEXIBLE
STEEL CONDUIT

ALUMINUM FLEXIBLE CONDUIT

Fig. 10-66 *Flexible-steel, aluminum, and liquid-tight conduit.*

FLEXIBLE LIQUID-TIGHT CONDUIT

This will damage the flex. On flexible conduit made of aluminum you will find the AL stamped every foot. This identifies it as aluminum. Flexible conduit should not be used in wet locations. However, it is permitted to be used if the conductors inside have a W designation. For example, THW or XHHW are types of conductors usable inside the flex in a wet location since they have a W designation.

If the W-type wire is used, it is also obvious that you should use watertight connectors. They will prevent water from running into equipment at the terminating end.

Supports

Flexible conduit requires support at much closer intervals than rigid conduit does. Supports should be every $4\frac{1}{2}$ feet and at least within 12 inches of a box or terminal. The bends made in the flex should be held rigidly so it will not deform when wires are pulled through it. There is an exception to this rule. A 5-foot length of $\frac{3}{8}$-inch flex can be used with two No. 14 AF wires with no grounding conductor and without support to supply power to a recessed incandescent lamp. Such a lamp is shown in Fig. 10-67.

Sealtite Flexible Cable

There are some uses for this type of waterproof flexible cable around the house. It can be used in wet locations and comes in sizes of $\frac{3}{8}$ inch to 4 inches. It does require a few special connectors for it to retain its waterproof conditions. These are shown in Fig. 10-68.

Fig. 10-67 *Using flexible conduit for connecting an incandescent lamp fixture in a conduit or a cable power supply.*

STCN connectors are made for type CN nonmetallic sealtite conduit. STCN means *sealtight connector, nonmetallic.* CN stands for *conduit, nonmetallic.* STCN connectors are made for use with Anaconda Type CN sealtite conduit (Fig. 10-69). (*Sealtite* is the usual spelling for *sealtight* conduit.) This type of conduit (CN) differs from standard metallic sealtight conduit in that it does not utilize a flexible metallic interior. Instead, CN conduit has a smooth plastic inner core, covered with a bonded reinforcing nylon cord. It also has a rugged plastic outer jacket for covering. The absence of any metallic elements in the conduit eliminates the problem of attack by corrosive atmospheres.

The STCN also has the advantage of making the conduit extra flexible. It provides an extremely smooth interior for protection of conductors. It is easily identified by its bright orange color. STCN forms a liquidtight and vaportight connection of CN conduit to a circuit breaker, switch, junction box, and similar enclosures, as well as to boxes and threaded hubs.

The STCN has four simple, easy-to-install parts: a nylon ferrule, a steel body, a compression nut, and a tiger-grip locknut (Fig. 10-70). The steel body of the connector has a built-in, protective plastic-insulated throat. It also has a neoprene O-ring.

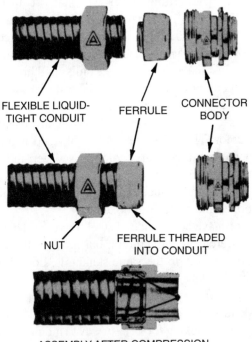

FLEXIBLE LIQUID-TIGHT CONDUIT

FERRULE

CONNECTOR BODY

NUT

FERRULE THREADED INTO CONDUIT

ASSEMBLY AFTER COMPRESSION SHOWING POSITIVE GROUND

Fig. 10-68 *Flexible conduit and connectors.*

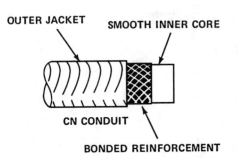

OUTER JACKET

SMOOTH INNER CORE

CN CONDUIT

BONDED REINFORCEMENT

Fig. 10-69 *Type CN sealtite conduit.*

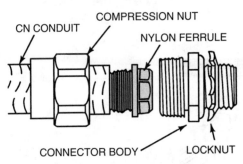

CN CONDUIT

COMPRESSION NUT

NYLON FERRULE

CONNECTOR BODY

LOCKNUT

Fig. 10-70 *Installing a type CN nonmetallic sealtite conduit conductor.*

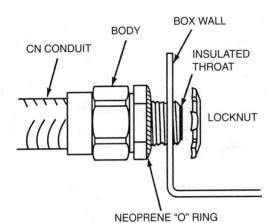

BODY

BOX WALL

INSULATED THROAT

CN CONDUIT

LOCKNUT

NEOPRENE "O" RING

Fig. 10-71 *Attaching CN conduit to a wall box.*

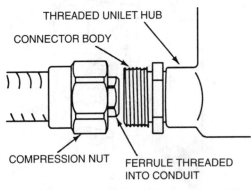

THREADED UNILET HUB

CONNECTOR BODY

COMPRESSION NUT

FERRULE THREADED INTO CONDUIT

Fig. 10-72 *Attaching the CN conduit to a unilet hub.*

To install the connector:

1. Cut the CN connector to the desired length, being sure to make a very straight cut.

2. Feed the compression nut onto the conduit.

3. Press the threaded end of the nylon ferrule as far as possible into the end of the conduit with the palm of the hand. Use the connector body as a wrench to tighten the ferrule flange snugly to the end of the conduit. (The interlocking hexagonal-nut type of design of the ferrule and body makes this possible.)

4. Bring the compression nut up and tighten to a positive step on the shoulder of the connector body.

5. Place the end of the body through the knockout and secure with the locknut provided (Fig. 10-71). Check Fig. 10-72 for installation of conduit into a threaded unilet hub.

Nonmetallic PVC

Nonmetallic PVC switch boxes and receptacle boxes are used in residential wiring. The roughing-in time can be reduced by using these boxes, inasmuch as they have their own connectors molded into the box. They can also be used in remodeling old work. See Fig. 10-73.

When you install an "old work" box into hollow walls, two arms swing out behind the wall. Tighten two screws, and the arms draw up tightly against the wall. Installation is completed with a few turns of

the screwdriver. A template is included in the box to ensure an accurate cut in the wall. The swing arm box can be used with wall material from $1/4$ to $5/8$ inch thick or from paneling to drywall. Just push the wire in and pull back slightly to activate the clamping action.

PVC boxes are made for switches and ceiling installations. The ceiling boxes will support up to 50 pounds. Mounting posts have a pair of holes $2^3/4$ and $3^1/2$ inches on center to accept fixture canopies that require either of these spacings. Heavy-duty nails are already inserted. Just make sure you don't miss and hit the box with a hammer—the results can be disastrous. The plastic does crack when hit a hard blow by a hammer that missed its target. The PVC boxes are usually a bright color. They may be gold or blue, to name but two of the manufacturers' colors. However, the thermoset (hard plastic) boxes are black. They also have raised covers and reducers. Figure 10-74 shows some PVC boxes and thermoset boxes.

Boxes have been made of steel for years. There are many different manufacturers, but most of the boxes are made roughly to the same standards. Look for the CSA and UL labels before you buy.

A wide variety of boxes are available. Each has its own identity and serves a specific purpose. Therefore, when you are looking for an easy way to mount a box, be sure to check all options. Many boxes come with brackets for specific types of mounting. See Fig. 10-75.

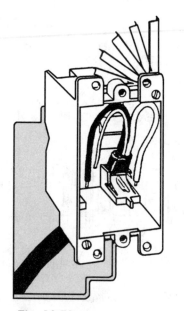

Fig. 10-73 *Nonmetallic box.*

3-GANG SWITCH BOX

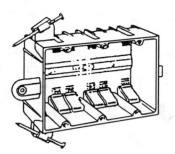

4" DIA. CEILING BOX

23.5 cubic inches

Fig. 10-74 *PVC boxes and thermoset boxes.*

On 4" square boxes clamps are furnished on two sides

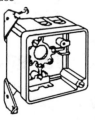

THERMOSET 4" SQUARE BOXES

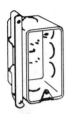

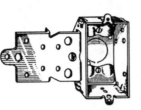

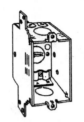

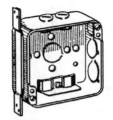

"A" BRACKET – Positions against side and face of stud. Bracket set back ¹/₂" on handy boxes.

"B" BRACKET – Mounts on face of stud. Bracket set flush on 4" square and set back ¹/₄" on handy and switch boxes.

"D" BRACKET – Flat, bracket side on Non-gangable switch boxes. Gauging notches at ¹/₈", ¹/₄" and ³/₈".

"FA" BRACKET – Side mount bracket with back-up flange. Set Back ¹/₂" on octagon and ¹/₈" on handy and switch boxes.

"FH" BRACKET – Side mount bracket with hooks that drive in face of stud. Flush on 4" square boxes and set back 4" on switch boxes.

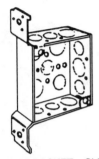

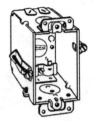

"FM" BRACKET – Side mount bracket for metal studs. Gauging notches set box ¹/₈" behind front of stud.

"J" BRACKET – Two spurs, slotted holes foe toe-nailing. Bracket set flush with gauging notches at ³/₈" and ¹/₂".

"MS" BRACKET – Snap-in bracket for metal studs. Mounts quickly without tools. Locks tightly in place for 1¹/₈", 2¹/₂", or 3³/₈" studs.

"S" BRACKET – Non-gangable switch box with sides extended. Option of staked, angled 16d nails. Gauging notches at ³/₈", ¹/₂" and ³/₈".

"T" BRACKET – Extension of sides of gangable switch boxes forms this T or tab bracket. Bracket useable even when boxes are ganged.

Fig. 10-75 *Boxes with brackets for different types of mounting.*

Plastic ears are included in many switch boxes. They are set forward ¹/₁₆ inch of the "old work" position. Two screw ears are supplied with shallow boxes and one screw ear on deep boxes. See Fig. 10-76. Clamps inside boxes are made for armored cable or nonmetallic cable (Romex). See Fig. 10-77.

Table 10-3 shows the volume needed per conductor so that the proper box can be chosen for a particular function. Remember that a clamp counts as a conductor.

Table 10-3 *Volume Required per Conductor*

Size of Conductor	Free Space within Box for Each Conductor
No. 14	2 cubic inches
No. 12	2.25 cubic inches
No. 10	2.5 cubic inches
No. 8	3 cubic inches
No. 6	5 cubic inches

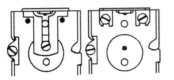

Fig. 10-76 *Boxes with screw ears for mounting. Left: one-screw ear; right: two-screw ear.*

CLAMPS

ARMORED CABLE

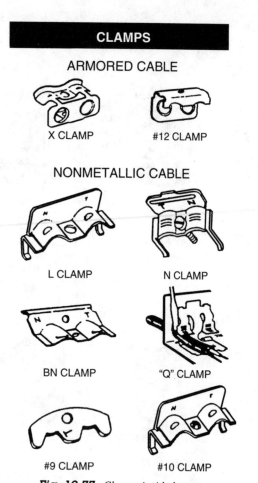

X CLAMP

#12 CLAMP

NONMETALLIC CABLE

L CLAMP

N CLAMP

BN CLAMP

"Q" CLAMP

#9 CLAMP

#10 CLAMP

Fig. 10-77 *Clamps inside boxes.*

11
CHAPTER

Remodeling Wiring

WHENEVER REMODELING IS CONSIDERED, ELECTRICAL requirements must be considered and given top priority. Each remodeling job is different. The electrician must be versatile and ingenious. Situations will arise that will not be covered in any book or any Code. Remember, though, that good wiring practices must be followed. The Code should be followed. However, since the variations created by the physical environment will alter the conventional methods, it may be difficult to follow the Code. It may take some ingenious methods, but in most cases, it can be done with a little thought and planning.

Needless to say, remodeling will take longer than new house wiring with drywall still to be installed. Each situation involves planning and the checking out of the old construction techniques.

Remodeling wiring usually calls for "adding on" to an existing system. The old system should be updated at this point, if necessary. That is, of course, if the owner is willing to provide the funds for such rewiring.

Figure 11-1 shows how to replace the older type of push-button switch with a more modern toggle-type of on/off switch. Note that the faceplate must also be changed. Before you replace the switch, a simple circuit tester should be used to determine if the circuit is live. A device such as that shown in Fig. 11-2 can be used. If the device glows when plugged into the outlet or across the proper terminals, it means the circuit is *live*, or *hot*.

CIRCUIT TESTING

Before replacing or adding onto an existing system, you must test the circuits to see if the wiring has been done correctly. This can be done in a number of ways. Figure 11-3 shows a circuit tester designed to locate

faults in wiring and to help check for continuity in existing outlet circuits. The circuit tester can be used to test circuits with outlets such as those shown in Fig. 11-4.

Fig. 11-2 *A fuse and circuit tester.*

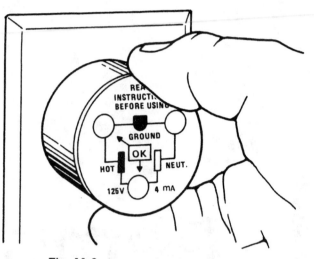

Fig. 11-3 *A receptacle polarity circuit tester.*

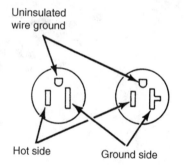

Fig. 11-4 *Two types of outlets with their markings.*

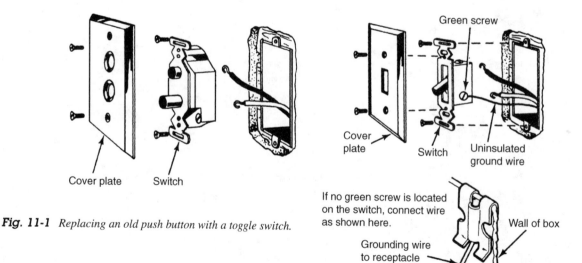

Fig. 11-1 *Replacing an old push button with a toggle switch.*

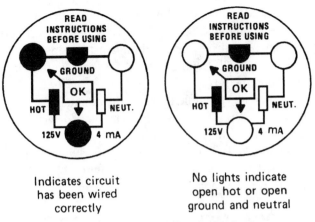

Indicates circuit has been wired correctly

No lights indicate open hot or open ground and neutral

Fig. 11-5 *Using the receptacle polarity circuit tester.*

By using an adapter, it is possible to check other types of circuits without the added neutral. This tester has a simple visual set of neon color-coded indicating lights to identify various fault conditions in electric circuits. The arrangement of lighted and/or unlighted indicating lamps will immediately show the circuit condition (Fig. 11- 5).

ADDING ANOTHER OUTLET

Switch and outlet boxes for additional outlets or switches must be located between studs. This will allow room for inserting the box and box supports. Decide where the box is to be located for the switch or the outlet. Mark the spot. Switches are located from 48 to 54 inches above the floor. They should be on the same level as existing switches. Convenience outlets are located about 12 to 18 inches above the floor. In the kitchen or dining area, the location may be different since they are needed for toasters and other appliances. That means they will be slightly above the dining table or kitchen working space.

Switches should be placed at the opening side of the door. Use the 2½-inch-deep box when possible.

Follow the steps shown in Fig. 11-6 for installing boxes. Use the sounding method—thumping on the wall—until you locate the studs. Some devices are available with magnets that point toward nail heads in the drywall or sheetrock, thus locating the stud.

Figure 11-6 shows the method used to install an additional box in old lath and plaster. Placing the box into a drywall installation is much simpler. Be sure to place a floor covering down to catch falling plaster or dust.

When you are cutting the hole in the wall with the hacksaw blade, hold the board or plaster with your hand to prevent cracking. After you complete the eight steps shown in Fig. 11-6, look at Fig. 11-7A. This shows one type of box to be used in this type of add-on wiring. This box has "grip-tight" brackets to hold the box in place. Figure 11-7B shows another type of box

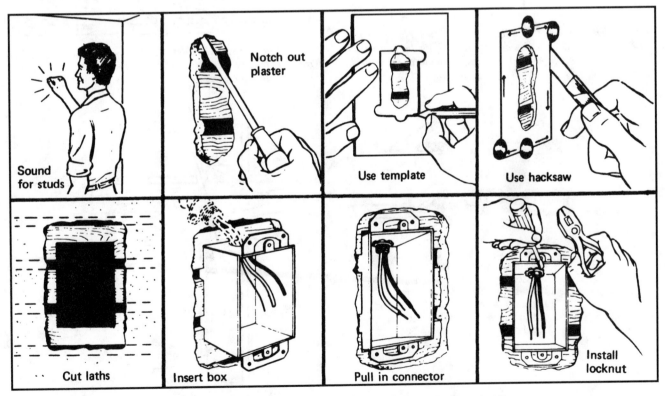

Fig. 11-6 *Steps in installing a box in an old building with plastered walls.*

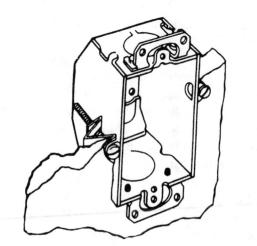

Fig. 11-7A *Using a box with grip-tight bracket to hold the box in place in the addition of a switch or outlet to existing wiring.*

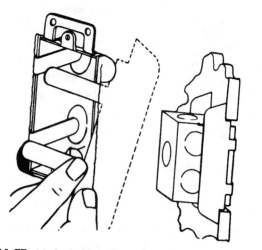

Fig. 11-7B *Method of installing a box with metal box supports. Insert the supports on each side of the box. Work the supports up and down until they fit firmly against the inside surface of the wall. Bend the projecting ears so that they fit around the box.*

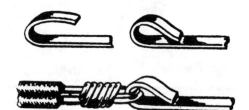

All fish tapes (or snakes) must
have a hook for fishing through conduit,
ceilings or partitions. This hook
is made by heating the end of the
fish tape on open flame until wire
becomes red hot and bending the end
of fish wire with pliers to correct the
size hook.

being used with metal box supports. Insert the supports on each side of the box. Work the supports up and down until they fit firmly against the inside surface of the wall. Bend the projecting ears so that they fit around the box.

When it is possible to fasten the box in place, look at Fig. 11-8. If either outlet is on an *interior* wall, you can usually drill straight upward to the partition between the walls. Secure the cable properly with staples as called for in your local code. Figure 11-9 shows how to run the cable across the basement.

WALL SWITCH ADDED TO CONTROL OUTLET

Most modern homes do not have overhead lamps in the bedrooms. They rely upon wall outlets to provide

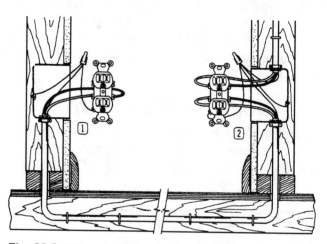

Fig. 11-9 *Wiring two outlets—extending one to another wall by means of the basement floor. Drilling through the floor is similar to the operation shown in Fig. 11-8.*

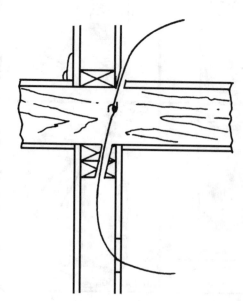

Fig. 11-8 *Fishing a wire with two fish wires.*

power for floor or table lamps. If a switch is needed to control a lamp, it is possible to add it without going through the basement (Fig. 11-10). This will be handy if you are in an upstairs bedroom and there is a finished ceiling below.

First, cut a hole in the wall. Remove the baseboard and cut holes B and C. Notch a channel in the plaster where the baseboard was originally. Remove the knockout in the switch box (A). Attach a cable connector. Fish the cable into the knockout in A, from B. Then run it through hole C to box D. Connect as shown. *Do not forget to attach the ground wire.* Attach it to the switch and the plug, if there is a screw for doing so. If not, attach it to the box with a screw or slip-on connector (Figs. 11-11 and Fig. 11-12). Replace the baseboard.

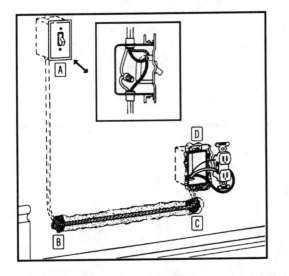

Fig. 11-10 *Removing the plaster between the two points and under the wallboard makes it easier to install a switch and outlet box.*

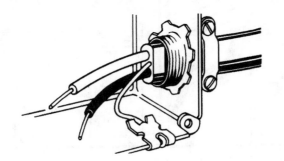

Fig. 11-11 *Romex connector attached to a box. Note the ground clip for the uninsulated ground wire.*

ADDING AN OUTLET ALONG THE WALL

Adding an outlet along the wall involves the procedure explained above. Figure 11-13 shows how an additional outlet is added to the wall. Extend a piece of wire from box E to box F. Place the cable in the groove cut for it

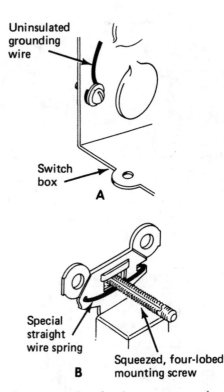

Fig. 11-12 *(A) Grounding the wire to a screw on the side of the metal box. (B) This is one method of attaching the ground wire to the box.*

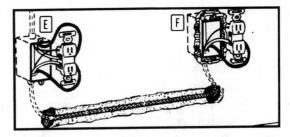

Fig. 11-13 *A wall-to-wall baseboard outlet extension.*

in the plaster, and replace the baseboard. Connect the wiring according to the color code. Connect black to black and white to white. Then run them to the new box. Make sure the black wire is connected to the brass-colored screw and the white wire to the silver-colored screw. Attach the ground wire (uninsulated) as shown in Fig. 11-11 or Fig. 11-12.

If you wish to place outlets in another room back to back with an existing outlet, look at Fig. 11-14. Use matching holes. Wire with conduit or threaded nipple with locknuts. Ground wires can be attached to the boxes with screws or clips.

RUNNING A WIRE AROUND A DOOR

Adding a new wall outlet may be difficult in some instances (Fig. 11-15). Remove the baseboard and the trim

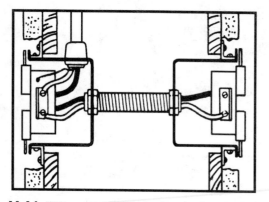

Fig. 11-14 *Wiring another box back to back with an existing outlet in another room.*

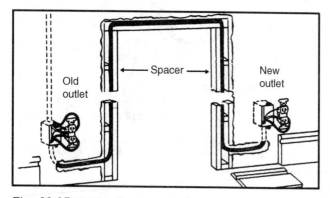

Fig. 11-15 *Running a wire around a door frame from the old outlet to a new one.*

around the door. Notch the wall and spacers between the frame and door jamb. If the outlet is to be placed beyond the first upright, use an extension bit to drill past the additional uprights. Wire the boxes and replace the baseboard and the door trim.

In some cases, headers are not always found in partitions (Fig. 11-16). They can be bypassed by using the method shown in Fig. 11-16A. They can also be notched, as shown in Fig. 11-16B. The hole can be patched and repainted, or wallpaper can be replaced carefully so as not to show the spot.

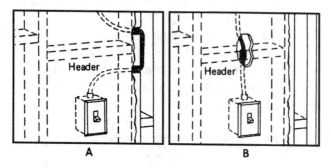

Fig. 11-16 *A. Running a cable around a header. B. Running a cable through a header.*

RUNNING A WIRE THROUGH THE FLOOR

In some cases, it is necessary to run a wire from one level of the house to another. As shown in Fig. 11-17, it is possible to add an outlet that is switched with the lamp. It is also possible to use an upstairs outlet and add a ceiling lamp in the basement or floor below. Select an outlet site and cut an opening through the lath and plaster.

If the outlet is on an outer wall, bore a diagonal hole from the basement. Use a long shank bit—if you are using a hand brace. Use a flat-type drill bit if you are using an electric hand drill. If the outlet is on the interior wall, bore through the floor to the partition between the walls. You do have to be careful to locate the wall and the partition 2 × 4. Use a piece of fish wire and push the cable up from the basement hole. Attach it to the wires and pull them through the outlet opening.

Attach the cable to the boxes as shown. Be sure to follow the correct color code for the outlet's brass and silver screws. Attach the extra uninsulated wires to the boxes with screws or clips.

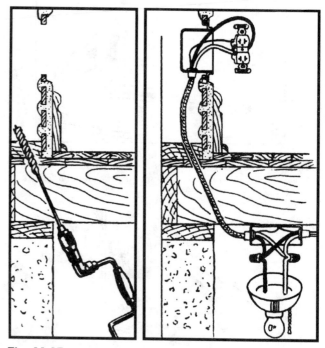

Fig. 11-17 *Running a wire through the floor from the basement to the upper floor.*

HOW TO FISH A WIRE

Figure 11-18 illustrates the method of using a fish wire to pull wire through holes drilled through the floor and wall. The fish wire comes in handy when a thermostat is relocated or a humidity control or an air conditioning thermostat is added.

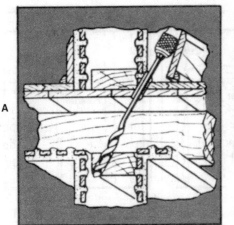

If you can get into attic or upper room, simply remove the upstairs baseboard. Then drill diagonal hole downward as shown.

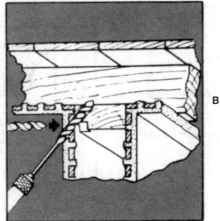

Drill diagonal hole upward from opposite room. Then drill horizontally until holes meet. This method requires patching plaster.

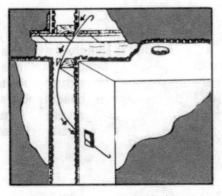

Push 12-foot fish wire, hooked at two ends, through hole on 2nd floor. Pull one end out of switch outlet on 1st floor.

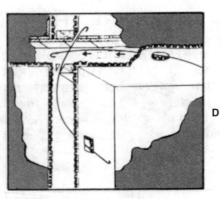

Next, push 20-25 foot fish wire, hooked at both ends through ceiling outlet (arrows). Now fish until you touch the first wire.

Fig. 11-18 *How to fish a wire.*

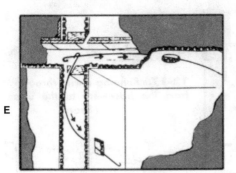

Then withdraw either wire (arrows) until it hooks the other wire; then withdraw second wire until both hooks hook together.

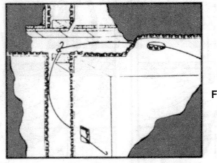

Lastly, pull shorter wire through switch outlet. When hook from long wire appears, attach cable and pull through wall and ceiling.

If you can get into an attic or upper room, simply remove the upstairs baseboard. Drill a diagonal hole downward as shown in Fig. 11-18A.

Drill the diagonal hole upward from the opposite hole. Then drill horizontally until the holes meet. This method requires patching the plaster (Fig. 11-18B).

Push the 12-foot fish wire, hooked at two ends, through the hole on the second floor. Pull one end out at the switch outlet on the first floor (Fig. 11-18C).

Push 20 to 25 feet of fish wire, hooked at both ends, through the ceiling outlet (note the arrows). Now fish until you touch the first wire (Fig. 11-18D).

Withdraw either wire (see the arrows) until it hooks the other wire. Then withdraw the second wire until both hooks come together (Fig. 11-18E).

Pull the shorter wire through the switch outlet. When the hook from the long wire appears, attach the cable and pull it through the wall and ceiling (Fig. 11-18F).

ATTIC INSTALLATION

The attic in older homes has floorboards. These can be carefully lifted and then replaced after the rewiring has been completed. Figure 11-19 shows how the floorboards can be lifted. The joists can be notched to hold the cable. A hole can be bored through obstructions with an electrician's bit. If the attic is inaccessible, but the cable can be run parallel with the floor beams or joists, connect the ceiling outlets with wall switches. Do this by drilling as shown in Fig. 11-20. Draw the cable through the opening, using fish wire. Where the cable runs across beams, the floorboards must be lifted. *When you reinstall the boards, be careful not to drive the nails through the cable.*

INSTALLING A SECURE BOX

In some instances, it is necessary to add a secure outside outlet. Keep in mind that outside outlets should have a ground fault circuit interrupter (GFCI). Providing the box and outlet with a key ensures that the plug is not used unless the person with the key is available.

Figure 11-21 shows how the adapter is fastened to a standard 4-inch or 4$\frac{11}{16}$-inch box. The side marked *top* is placed at the top of the box. Run the conduit to the box as usual in the case of a brick wall. It is possible to use Romex in regular frame construction. Set the box in the wall so the front edge of the adapter will be flush, or recessed not more than $\frac{1}{4}$ inch from the finished wall.

Install the finish frame and cover after applying caulking compound to the back of the frame to provide

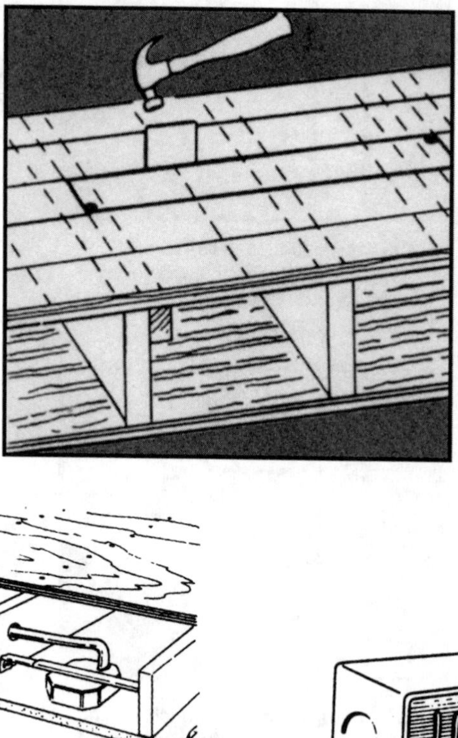

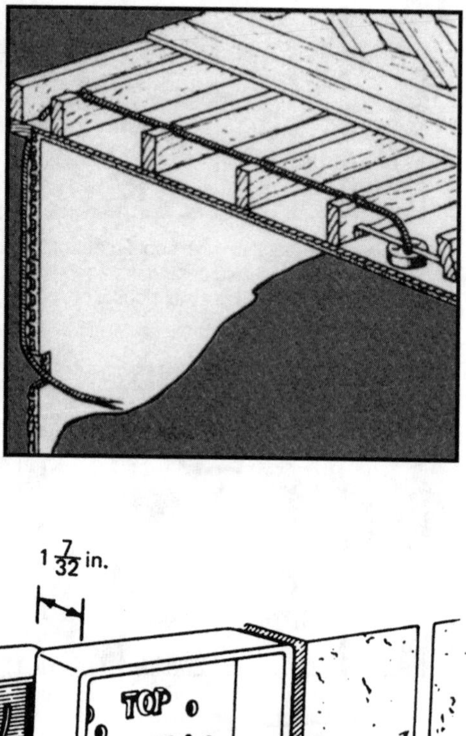

Fig. 11-19 *Lifting the floor of an attic to get at the ceiling for purposes of wiring an outlet or lamp.*

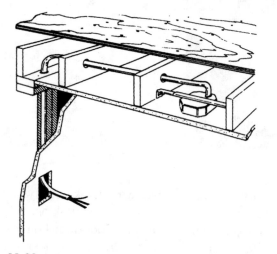

Fig. 11-20 *Lift the floorboards and place the wires in a notch, or drill through the rafters or joists and feed the wire through the holes. Be careful in reinstalling the floorboards to make sure the nails do not penetrate the cable.*

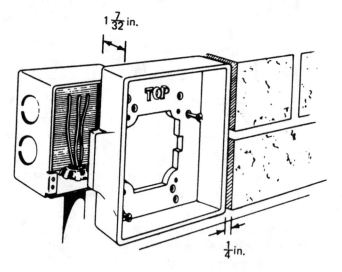

Fig. 11-21 *Installing a secure type of outlet for outdoor use.*

a weatherproof seal (Fig. 11-22). Elongated slots in the frame allow for alignment.

Wire the device used—in this case the duplex outlet—and install it in the box. Note that the device must have a special mounting plate (Fig. 11-23). The 30- and 50-ampere receptacles should be installed with the ground contacts at the top to facilitate the use of angle cord sets.

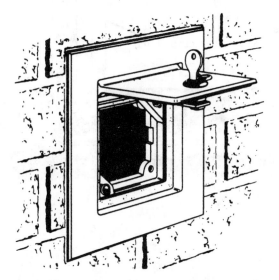

Fig. 11-22 *Place the finish frame over the square box.*

Fig. 11-23 *Install the GFCI or outlet.*

Other Outside Switches and Receptacles

The switches and receptacles used outside must be weatherproof. Weatherproof devices must be installed where motors and high-current farm equipment are plugged into the circuit. These switches and receptacles vary in size (Fig. 11-24). Keep in mind that a motor

with a rating of more than ½ horsepower (373 watts) must have 240 volts to supply it properly. This will require three-wire installation and types of plugs different from those used on 120-volt. One-horsepower (746- watt) motors draw about 6.5 amperes on a 115-volt line. This means there will be little current for other appliances or lights on the 15-ampere line. Therefore, it is best to use the three-wire, 240-volt line for motors. Remember, there is some voltage drop along the wire. This is especially true if the distance from the circuit breaker box is more than 25 feet.

When you are wiring the outside boxes, make sure the black wire goes to the brass-colored screw on the receptacle. If a red wire is used, it goes to the brass-colored screw since it, too, is a hot line. The white goes to the silver-colored screw on a receptacle.

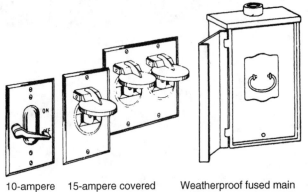

| 10-ampere toggle switch | 15-ampere covered receptacles. Choice of single or duplex types. | Weatherproof fused main switch for mounting outdoors on pole or building wall |

Fig. 11-24 *Weatherproof switches and outlets for outdoor use.*

UNDERGROUND WIRING

You may want to add a light or garage door opener to a garage that is located some distance from the house. This can be done by placing the wiring underground (Fig. 11-25).

Check your local code to see which type of wire is approved for wiring underground. You may want to use an installation similar to that shown in Fig. 11-25. Here, conduit is used with waterproof fittings. If you do not care for the extra expense of this type of wiring, you can substitute (if your code allows) the underground cable that resembles Romex.

By using the plastic-coated wiring, it is possible to link houses and outbuildings easily (Fig. 11-26). Farms can use this type of wiring easily and bury it. By burying the wire, it is protected from weather, vehicle traffic, and other hazards. In some localities, you are required by local code to use only conduit. Check to make sure before you install plastic cable. Plastic cable can also be used to run through concrete or masonry.

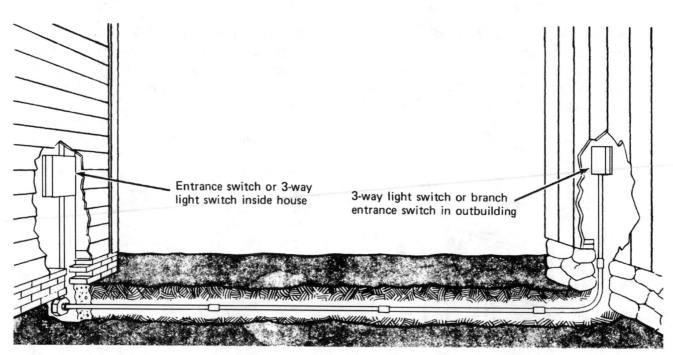

Fig. 11-25 *Wiring underground with weatherproof conduit.*

Entrance switch or 3-way light switch inside house

3-way light switch or branch entrance switch in outbuilding

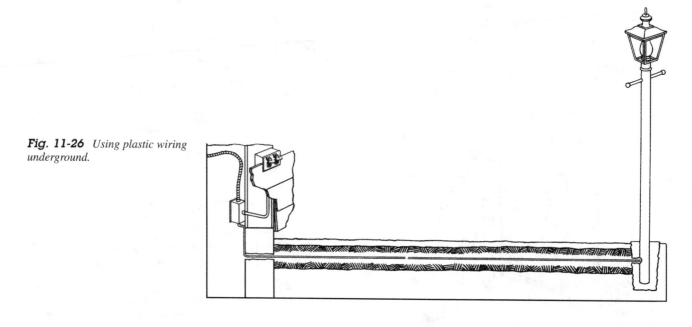

Fig. 11-26 *Using plastic wiring underground.*

Today's wire is treated so that rodents will not chew on it when it is buried.

The National Electrical Code allows the use of Type UF plastic wire for burying underground. It comes in No. 14 through 4/0 for use as a branch circuit and underground feeders. It should be buried no less than 12 inches if it is used for residential circuits with less than 300 volts and no more than 30 amperes.

Wet and Damp Locations

In most instances, underground wiring will terminate in an outlet. This outlet may be a receptacle, switch, or fixture. Installation of a receptacle in a damp location is governed by the Code. An open porch or a screened-in porch is a damp location. A conventional receptacle cannot be used in this location. The receptacle must have a cover plate that is weatherproof. The UL lists plates according to use, such as "weatherproof." The cover plate may have a screw-on cap such as that shown in Fig. 11-27. Figure 11-28 shows the side-opening type of covers for a receptacle. If the doors stay open, they are suitable for use in damp areas, such as porches. If such covers are self-closing, they can be used outside to supply portable tools or equipment.

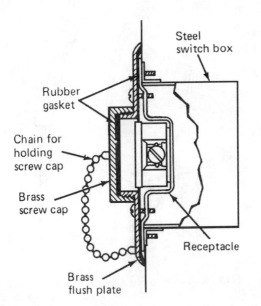

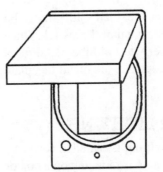

Fig. 11-27 *A chain-held screw-cap cover for damp locations.*

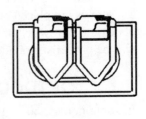

Fig. 11-29 *This swing-down canopy is self-closing. The duplex cover is also self-closing.*

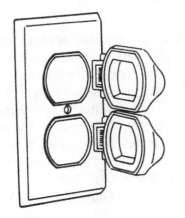

Fig. 11-28 *Side-opening covers. If the doors are self-closing, such a cover assembly could be used outside. It could be used to supply portable tools if the equipment is not left connected.*

A receptacle used in a wet area must be kept waterproof at all times when the plug is inserted and when it is removed.

Figure 11-29 shows a swing-down, self-closing canopy. It keeps the area shielded from rain when the receptacle is in use or plugged in. A duplex cover is also available for outdoor use. The NEC also requires an elevation of outdoor receptacles to prevent accumulation of water.

COVERS AND CANOPIES

The NEC requires (and it is common sense to do so) that every outlet box be covered by a cover plate, a fixture canopy, or a faceplate. The openings for a receptacle, snap switch, or other device installed in the box must be covered.

One part of the Code requires all metal faceplates to be grounded. Because metal faceplates are *exposed*

conductive parts, they must meet Code requirements. This means that ungrounded metal faceplates shall not be installed within 8 feet vertically or 5 feet horizontally of ground or grounded objects. This refers to laundry tubs, bathtubs, showers, plumbing fixtures, steam pipes, radiators, or other grounded surfaces that may be contacted by people.

Another part of the Code requires metal faceplates to be grounded in wet or damp locations, unless the faceplates are isolated from contact. Such locations can include bathrooms. It is a good idea not to install metal faceplates on switches or receptacles in bathrooms or kitchens. They may be hazardous if the home does not have the proper grounding system. If a nonmetallic box is used for the receptacle or switch, the grounding type of switch must be used (Fig. 11-30). For any metal faceplate on nonmetallic boxes, the

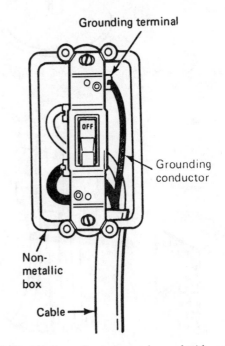

Fig. 11-30 *This type of switch must be used with metal switch plates mounted on nonmetallic boxes.*

grounding conductor from the feed cable must be attached to the grounding terminal of the switch. Once the screws are placed properly to hold the metal faceplate onto the switch or receptacle, the metal faceplate will be properly grounded.

OUTDOOR LIGHTING

In some homes, outdoor lighting is installed for year-round use. This means the lamp is mounted on a box or conduit-fed fixture. Some precautions should be kept in mind under these conditions. A single rigid metal conduit may not support a box, even with concrete fill in the ground (Fig. 11-31). The box should be supported by two metal conduits. EMT cannot be used to support a box. It should be supported by rigid conduit. Figure 11-32 shows a number of EMTs supporting the box. This is not according to the Code, even though several connections are made to the box. Figure 11-33 shows

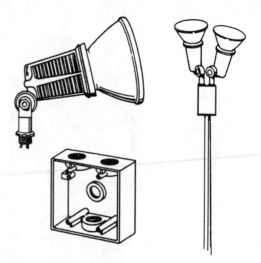

Fig. 11-33 *Supporting a box with EMT is not according to the Code. Having the box over 18 inches aboveground is not according to the Code. Supporting the box with only one piece of conduit is also a violation of the Code.*

three of the common violations of the NEC. The box shown is supported by a piece of EMT—violation number one. Only one hub on the box is connected—violation number two. The box is more than 18 inches above the ground—violation number three.

Any outlet box that is mounted on conduit must not be over 18 inches above ground. It should have two pieces of conduit supporting it by connecting to at least two hubs on the box. EMT should not be used to support a box. It does not have the strength to withstand severe blows.

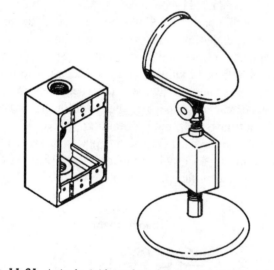

Fig. 11-31 *A single rigid metal conduit cannot be used to support a box, even if mounted in concrete.*

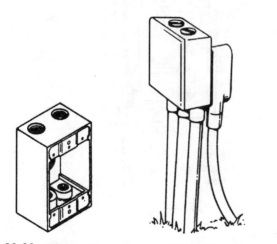

Fig. 11-32 *The use of more than two EMT supports per box is not according to the Code.*

12
CHAPTER

Appliances, Ovens, Ranges, Cooking Tops, Air Conditioners, Cable TV, and Internet

ANY MODERN APPLIANCES ARE PORTABLE AND use a standard 120-volt circuit with at least 15 amperes. However, some devices use 240 volts and need special plugs and wiring.

ELECTRIC RANGES

The largest appliance in the house is usually an electric range for cooking. It is wired with a permanent connection or with a plug and pigtails. Figure 12-1 shows the plug that connects the range terminals provided by the manufacturer to the surface-mounted box in Fig. 12-2. This plug and receptacle must be capable of handling at least 50 amperes and provide connections for the three wires used in such circuits.

Figure 12-3 shows a flush-mounted outlet for 240 volts. It can be wired in a junction box with the wall plate to fit (Fig. 12-4). In Fig. 12-5 the dryer is shown hooked up properly with a pigtail flexible cord. This type of setup can be used for connecting electric

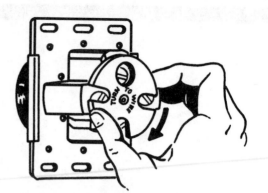

Fig. 12-3 *This outlet must be turned to wire.*

Fig. 12-4 *A faceplate for a flush-mounted single receptacle.*

Fig. 12-1 *A pigtail and plug.*

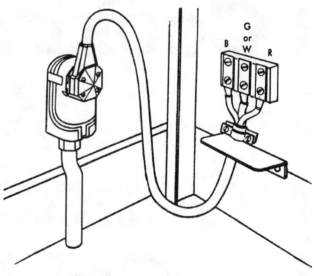

Fig. 12-5 *Installation of a pigtail.*

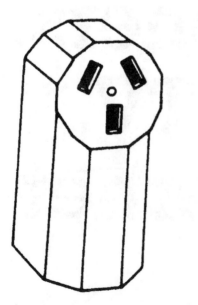

Fig. 12-2 *A surface-mounted receptacle for 240 volts, 50 amperes.*

ranges. The 50-ampere line will be No. 6 wire with red, black, and white leads. The green lead is a ground wire that is connected to the metal part of the dryer, range, or heating unit for the cooking surface.

The plug and outlet shown in Fig. 12-5 are allowed only for convenience. They cannot serve as a switch or as a means of disconnecting the appliance from the line. A separate switching method is necessary to disconnect the circuit from the oven, cooking top, or dryer.

All new houses are built with a dishwasher and a garbage disposal. These appliances are usually prewired (come with a power cord) and can be plugged into a wall receptacle. Check thses appliances to make sure their current demands are not more than rated for the circuit. On/off switches for them should be installed on the wall near the kitchen sink.

Connecting Ranges Permanently

The National Electrical Code specifies the size of wire that can be used for connecting electric ranges. Feeder capacity must be allowed for household cooking appliances according to Table 220-19 of the Code. This table applies if the appliance is rated over 1¾ kilowatts.

Section 210-19(b), Exception No. 1, refers to Table 220-19 for the sizing of a branch circuit to supply an electric range, a wall-mounted oven, or a counter-mounted cooking unit. Table 220-19 is also used for sizing the feeder wires that supply more than one electric range or cooking unit.

Figure 12-6 shows that the minimum size for a 12,000-watt range would be 35 amperes, and the wire would have to be No. 8. The type of wire would be TW–40 A, THW–45 A, XHHW, or THHN–50 A. Number 8 wire is the minimum size to be used for any range rated 8¾ kilowatts or more. The overload protection could be either a 40-ampere fuse or a 40-ampere circuit breaker.

Although the two hot legs of the wire must be No. 8, the exception to the Code says that the neutral conductor can be smaller. See Fig. 12-6. This is so primarily because the heating elements of the range will be connected directly between the two hot wires and the maximum current will flow through these wires. The devices that operate on 115 volts will be the oven lamp and the lighting circuits over the top of the range. In some instances, the heating elements for the top portion will be wired to 115 volts at some heat levels.

The exception says that the smaller conductor can be not less than 70% of the capacity of the hot wires. In this case, the wire cannot be less than No. 10. The neutral size is figured as follows: 70% of 40 amperes is 28 amperes. Thus, 28 amperes calls for No. 10 wire. The neutral can be No. 10, but the other two wires must be No. 8.

If the range is over 8¾ kilowatts, then the minimum size of the neutral conductor must not be smaller than No. 10. Hot legs of the range must not be less than No. 8 wire and the neutral less than No. 10 wire.

The maximum demand for a range of 12-kilowatt rating according to the table in the Code is 8 kilowatts. This 8000-watt load can be converted to amperes by dividing by 230 volts. This, then, gives 35 amperes. Look up the wire for handling 35 amperes, and you find No. 8. Number 8 wire can handle up to 40 amperes.

On modern ranges, the heating elements of the surface units are controlled by five-heat switches. The surface unit heating elements will not draw current from the neutral conductor unless the unit switch is in one of the low-heat settings.

Sizing a Range over 12 Kilowatts

If you have a kitchen range rated at 16.6 kilowatts, then the 230-volt supply must be capable of delivering at least 43 amperes. This is determined by the following:

1. Column A of the NEC Table for *Demand Loads for Household Ranges, Wall-Mounted Ovens, Counter-Mounted Cooking Units and Other Household Appliances Over 1¾ kW Rating* states in Note 1 that for ranges over 12 kilowatts and up to 27 kilowatts, the maximum demand in column A must be

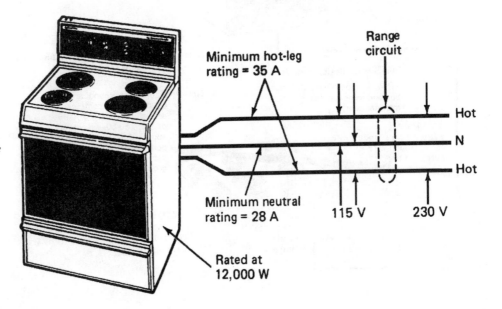

Fig. 12-6 A 12-kilowatt range with its minimum ampere rating.

Minimum hot-leg rating = 35 A

Range circuit

Hot

N

Hot

Minimum neutral rating = 28 A

115 V   230 V

Rated at 12,000 W

increased by 5% for each additional kilowatt (or major fraction thereof) above 12 kilowatts.

2. This range is rated at 16.6 kilowatts, which is 4.6 kilowatts more than 12 kilowatts. Column A says to use 5% of the 8000 watts, or 400 watts.

3. The maximum demand for this 16.6-kilowatt range must be increased above 8 kilowatts by using the following: 400 watts × 5 (4 kilowatts + 1 for the 0.6 kilowatt more than 16). Therefore, 400 × 5 = 2000. This is added to the 8000 watts used as a base demand and produces 10,000 as the maximum demand load.

4. If you want the amount of current drawn at maximum demand, divide 10,000 watts by 230 volts ($P = E \times I$, or $I = P \div E$). The answer is 43.478 amperes. This is rounded to 43 since the fraction is less than 0.5. Look up the wire size for using a 60°C conductor, and you will find the required UL wire for a branch circuit must have No. 6 TW conductors to handle 43 amperes.

Tap Conductors

Section 210-19(b), Exception 1, of the NEC gives permission to reduce the size of the neutral conductor for ranges or a three-wire range branch circuit to 70% of the current-carrying capacity of the ungrounded conductors. Keep in mind, however, that this does not apply to smaller taps connected to a 50-ampere circuit where the smaller taps must all be the same size. It does not apply when individual branch circuits supply each wall- or counter-mounted cooking unit, and all circuit conductors are of the same size and less than No. 10.

Exception No. 2 of the section previously mentioned allows tap conductors rated at not less than 20 amperes to be connected to a 50-ampere branch circuit that supplies ranges, wall-mounted ovens, and counter-mounted cooking units. These taps cannot be any longer than necessary for servicing the cooking top or oven. See Figs. 12-7 and 12-8.

Figure 12-9 shows two units treated as a single range load. Figure 12-10 shows how to determine the branch circuit load for separate cooking appliances on a single circuit according to Section 210 of the Code.

Individual branch circuits can be used. There are some advantages to individual circuits. With this arrangement, smaller branch circuits supply each unit with no junction boxes required. Two additional fuses or circuit breaker poles are required in a panelboard, however. Overall labor and material costs are less than those for the 50-ampere circuit shown in Fig. 12-8. There is a disadvantage though. The smaller circuits will not handle larger units that you may wish to install later. See Fig. 12-11.

In some cases, a single 40-ampere circuit may supply the units (Fig. 12-12). The NEC allows 40-ampere circuits in place of 50-ampere circuits where the total

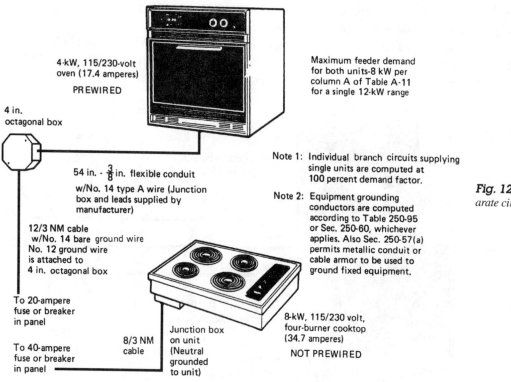

4-kW, 115/230-volt oven (17.4 amperes) PREWIRED

4 in. octagonal box

54 in. - ⅜ in. flexible conduit w/No. 14 type A wire (Junction box and leads supplied by manufacturer)

12/3 NM cable w/No. 14 bare ground wire No. 12 ground wire is attached to 4 in. octagonal box

To 20-ampere fuse or breaker in panel

To 40-ampere fuse or breaker in panel

8/3 NM cable

Junction box on unit (Neutral grounded to unit)

8-kW, 115/230 volt, four-burner cooktop (34.7 amperes) NOT PREWIRED

Maximum feeder demand for both units-8 kW per column A of Table A-11 for a single 12-kW range

Note 1: Individual branch circuits supplying single units are computed at 100 percent demand factor.

Note 2: Equipment grounding conductors are computed according to Table 250-95 or Sec. 250-60, whichever applies. Also Sec. 250-57(a) permits metallic conduit or cable armor to be used to ground fixed equipment.

Fig. 12-7 *Cooking units with separate circuits.*

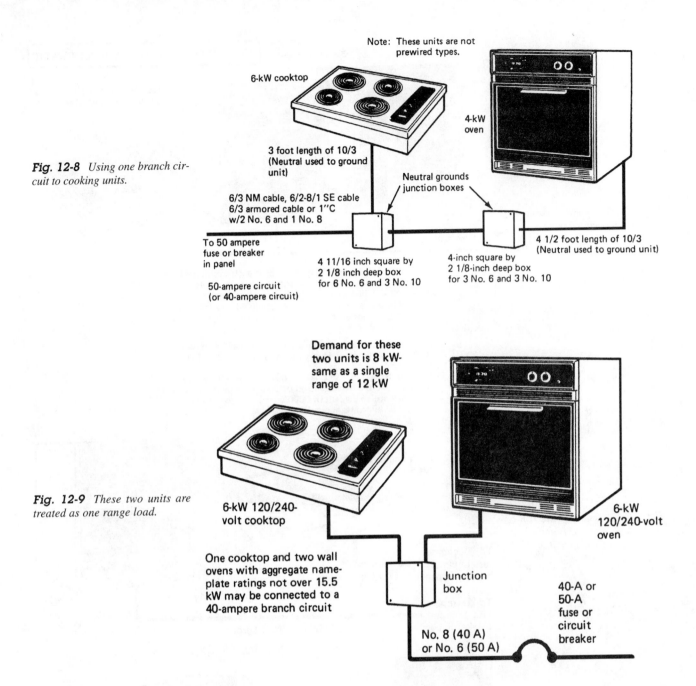

Fig. 12-8 *Using one branch circuit to cooking units.*

Note: These units are not prewired types.

6-kW cooktop

4-kW oven

3 foot length of 10/3 (Neutral used to ground unit)

Neutral grounds junction boxes

6/3 NM cable, 6/2-8/1 SE cable 6/3 armored cable or 1''C w/2 No. 6 and 1 No. 8

To 50 ampere fuse or breaker in panel

50-ampere circuit (or 40-ampere circuit)

4 11/16 inch square by 2 1/8 inch deep box for 6 No. 6 and 3 No. 10

4-inch square by 2 1/8-inch deep box for 3 No. 6 and 3 No. 10

4 1/2 foot length of 10/3 (Neutral used to ground unit)

Fig. 12-9 *These two units are treated as one range load.*

Demand for these two units is 8 kW-same as a single range of 12 kW

6-kW 120/240-volt cooktop

6-kW 120/240-volt oven

One cooktop and two wall ovens with aggregate nameplate ratings not over 15.5 kW may be connected to a 40-ampere branch circuit

Junction box

40-A or 50-A fuse or circuit breaker

No. 8 (40 A) or No. 6 (50 A)

nameplate rating of the cooktops and ovens is less than 15.5 kilowatts. Since most ranges and combinations of cooktops and ovens are less than 15.5 kilowatts, this can be a very popular arrangement.

DRYER

The electric clothes dryer uses 240 volts for the heating element. It also uses 120 volts for the motor and 120 volts for the lightbulb. This means three wires are needed for proper operation. Figure 12-13 shows how the green and white wires are treated in the dryer. Since most home dryers draw about 4200 watts, it is necessary to use a separate 30-ampere circuit with a pullout fuse for disconnecting it from the line. A circuit breaker in the main service panel can also be used to disconnect it from the line. The 30-ampere circuit uses No. 8 wire in most instances. However, note that some high-speed dryers use about 8500 watts and need a 50-ampere circuit. That would call for a No. 6 wire. There is a difference in the configuration of the receptacle for a 50-ampere device and that for a 30-ampere device (Fig. 12-14). The surface-mounted receptacle with an L-shaped ground is shown in Fig. 12-14A. It is used with dryers drawing up to 30 amperes. Figure 12-14B shows a surface-mounted receptacle also. It is capable of handling up to 50

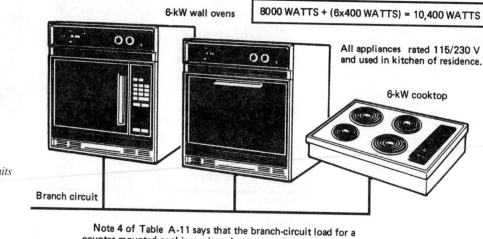

6-kW wall ovens

$$8000 \text{ WATTS} + (6 \times 400 \text{ WATTS}) = 10,400 \text{ WATTS}$$

All appliances rated 115/230 V and used in kitchen of residence.

6-kW cooktop

Fig. 12-10 *Separate cooking units on a single circuit branch.*

Branch circuit

Note 4 of Table A-11 says that the branch-circuit load for a counter-mounted cooking unit and not more than two wall-mounted ovens, all supplied from a single branch circuit and located in the same room, shall be computed by adding the nameplate ratings of the individual appliances and treating this total as a single range.

That means the three appliances shown may be considered to be a single range of 18-kW rating (6 kW + 6 kW + 6 kW).

From Note 1 of Table A-11, such a range exceeds 12 kW by 6 kW and the 8-kW demand of Column A must be increased by 400 watts (5 percent of 8000 watts) for each of the 6 additional kilowatts above 12 kW.

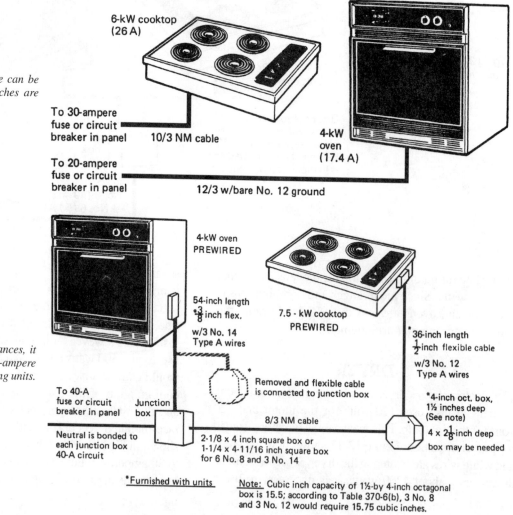

6-kW cooktop (26 A)

Fig. 12-11 *Smaller wire can be used when separate branches are run for each cooking unit.*

To 30-ampere fuse or circuit breaker in panel   10/3 NM cable

4-kW oven (17.4 A)

To 20-ampere fuse or circuit breaker in panel

12/3 w/bare No. 12 ground

4-kW oven PREWIRED

54-inch length $\frac{3}{8}$ inch flex. w/3 No. 14 Type A wires

7.5 - kW cooktop PREWIRED

Fig. 12-12 *In some instances, it is possible for a single 40-ampere circuit to supply both cooking units.*

Removed and flexible cable is connected to junction box

36-inch length $\frac{1}{2}$ inch flexible cable

w/3 No. 12 Type A wires

To 40-A fuse or circuit breaker in panel   Junction box

8/3 NM cable

*4-inch oct. box, 1½ inches deep (See note)

Neutral is bonded to each junction box 40-A circuit

2-1/8 x 4 inch square box or 1-1/4 x 4-11/16 inch square box for 6 No. 8 and 3 No. 14

4 x 2$\frac{1}{8}$-inch deep box may be needed

*Furnished with units

Note: Cubic inch capacity of 1½-by-4-inch octagonal box is 15.5; according to Table 370-6(b), 3 No. 8 and 3 No. 12 would require 15.75 cubic inches.

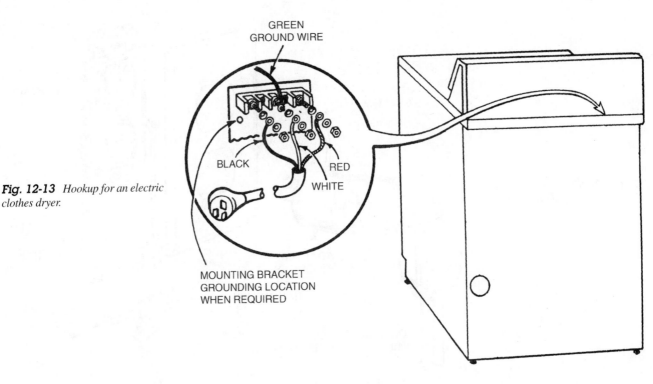

Fig. 12-13 *Hookup for an electric clothes dryer.*

GREEN GROUND WIRE

BLACK

RED

WHITE

MOUNTING BRACKET GROUNDING LOCATION WHEN REQUIRED

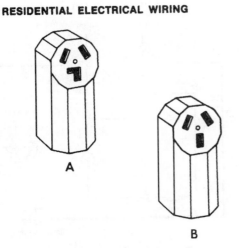

RESIDENTIAL ELECTRICAL WIRING

A

B

Fig. 12-14 *(A) L-type ground for a 250-volt, 30-ampere receptacle. (B) 250-volt, 50-ampere range receptacle.*

amperes at 240 volts. It is used for dryers or ranges. Note the difference in the three outlet connection points. This one has three straight slots.

MICROWAVE OVENS

Microwave ovens are designed to operate on 120 volts. This means they can be plugged into the closest convenient wall outlet. This may become an expensive mistake for many. The microwave oven pulls around 14 amperes. This means there is only 1 ampere of spare capacity left for the circuit. If there is another appliance on the circuit, a circuit breaker or fuse may go. Microwave ovens should have their own circuit. There should be nothing

else on the line. They should have a direct connection to the distribution panel. Make sure they are properly grounded as shown in Figs. 12-15 and 12-16.

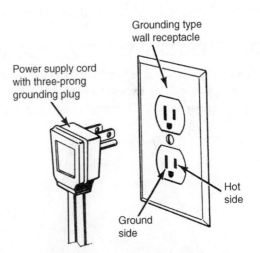

Grounding type wall receptacle

Power supply cord with three-prong grounding plug

Hot side

Ground side

Fig. 12-15 *A single-phase, 120-volt plug and receptacle.*

ELECTRIC HEATERS

There are a number of different designs available to the homeowner today in the field of electric heaters. Heaters such as that shown in Fig. 12-17 may be effective and serve their purpose, but they draw more current than most owners care to admit. They should have a separate circuit to furnish the 120 volts and almost 15 amperes. They have a small electric motor for moving the air, a lightbulb for the simulated logs, and a small motor for

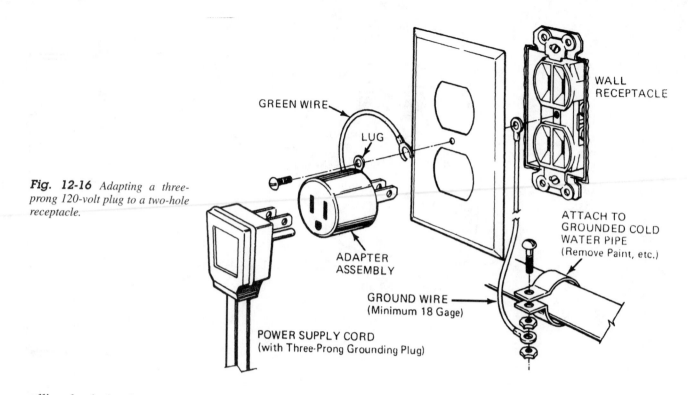

Fig. 12-16 *Adapting a three-prong 120-volt plug to a two-hole receptacle.*

pulling the device that makes the logs resemble a smoldering fire. The combination of circuitry is shown in Fig. 12-18. Other small heaters should be checked to see how much current they draw. Take the wattage rating and divide by 120 to obtain the current draw of the device.

ROOF HEATERS AND PIPE HEATERS

In the colder climates it is necessary to heat the eaves to prevent water from backing up under the roofing when the snow melts. These heating elements are

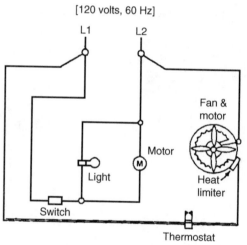

Fig. 12-18 *A circuit for Fig. 12-17.*

Fig. 12-17 *A wall-mounted, decorative electric heater.*

installed as shown in Fig. 12-19. Note that the manufacturer has suggested that the heating element be plugged into an outlet such as that servicing the lamp. This is permissible when the heating element is not as long as that shown in Fig. 12-20. The length of the cable determines its wattage rating and, in effect, the current drawn. A separate circuit may be necessary for such heating elements if the current drawn is more than that of the present circuit.

Take a look at Tables 12-1 and 12-2 to see at what point a new line may be necessary for operation of this heating element. Also keep in mind that adding a line would call for waterproof outlets and proper connection of the cord to the circuit inside a waterproof box.

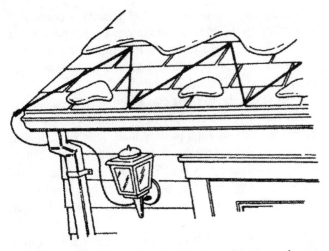

Fig. 12-19 *Installation of roof heaters for melting ice and snow.*

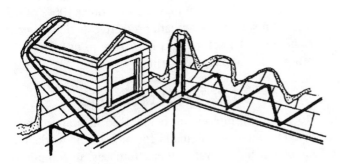

Fig. 12-20 *Installation of roof heaters around a dormer.*

As can be seen by comparing Tables 12-1 and 12-2, it is better to have the 240-volt line supplying power to the heating element. This would, of course, necessitate a separate line to supply the demand. However, if this is known when the house is built, the line can be added then, rather than later, after construction is complete. Heating elements for water lines exposed to cold temperatures also use the heat produced to prevent frozen and cracked pipes.

Table 12-1 *110- to 120-Volt Heating Elements*

Length of Cable (feet)	Average Length of Roof Edge (feet)	Wattage	Current (Amperes)
20	10	120	1.0
30	15	180	1.5
40	20	240	2.0
60	30	360	3.0
80	40	480	4.0
100	50	600	5.0
120	60	720	6.0
140	70	840	7.0
160	80	960	8.0

Table 12-2 *220- to 240-Volt Heating Elements*

Length of Cable (feet)	Average Length of Roof Edge (feet)	Wattage	Current (Amperes)
40	20	240	1.0
60	30	360	1.5
80	40	480	2.0
120	60	720	3.0
160	80	960	4.0
200	100	1200	5.0

OVERHEAD GARAGE DOORS

Most overhead garage doors for residential use are furnished with a $1/3$-horsepower (248.6-watt) electric motor. This means it draws about 2.07 amperes when running. It does, however, draw much more when starting. This is what dims the lights when it starts. Lightbulbs and other electrical devices are sensitive to some high peak voltages.

Every time the garage door motor turns off, it gives an inductive kickback. This increased voltage spike produces a shorter life for lightbulbs on the same circuit.

Most manufacturers of garage door openers stress that they are easy to install and can be plugged into the existing lamp socket in the garage (Fig. 12-21). It is

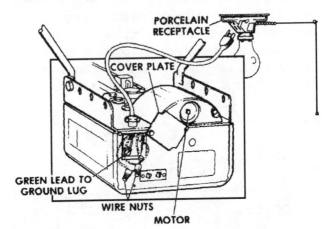

Fig. 12-21 *Wiring of an overhead garage door opener.*

suggested, however, that the plug connection to the box shown in Fig. 12-22 be substituted and Romex from a direct line be placed in the garage for operation of the garage door opener.

ELECTRIC WATER HEATERS

Electric water heaters draw current to heat the water. The larger the heater capacity, the higher the current rating of the heating element. They use 240 volts for heating the water.

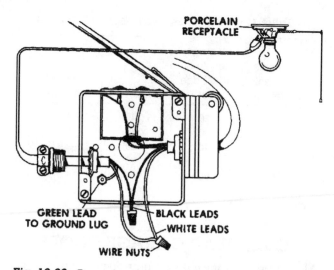

Fig. 12-22 *Connection of a garage door opener to the power line.*

Labels in figure: PORCELAIN RECEPTACLE; GREEN LEAD TO GROUND LUG; BLACK LEADS; WHITE LEADS; WIRE NUTS

A double-element water heater is probably better for larger families because it permits a more constant supply of hot water. Double-element heaters have two thermostats. The single-element type has only one thermostat. The size of the elements, the type of thermostats, and the method of wiring for heaters are usually specified by the local power company.

Figure 12-23 shows a typical installation of a hot water heater. Keep in mind that some electric companies offer a special rate for heating water. *Off-peak load* is the term that refers to this special rate. It simply means the power company furnishes electricity during its low-load time. The time is already known by the electric company so it can place a meter that measures the use of power for heating water. It also places a time switch

on the line so that the usage of electricity is controlled or limited to the time when the power company has a very light demand for power. If you use power at any time other than off-peak time, you have to pay a standard rate, instead of the reduced rate.

As far as the Code is concerned, any fixed storage water heater with a capacity of 120 gallons or less must be treated as a continuous-duty load. This will apply to about 90% of the residential water heaters that use electricity for heating water. The *continuous-duty load* means the ampere rating of the water heater must not exceed 80% of the ampere rating of the branch circuit conductors.

The only case where the water heater current may load the circuit protective device (circuit breaker or fuse) to 100% of its rating is where the circuit protective device is listed for continuous operation at 100% of its rating. Presently, no standard protective device is rated in this way. Therefore, it is necessary to use the 80% rating as a guide to wire sizing.

ELECTRIC MOTORS

More electric motors are being used in homes every day. The increased demand for motors for workshop equipment and for water pumps, furnaces, air conditioners, and other devices creates a need for a closer look at circuits that furnish the power for electric motors. Table 12-3 shows the wire sizes for individual single-phase motors. It should help with the selection of the proper wire size for workshop use or any other motor application around a residence (Fig. 12-24).

Garbage Disposers

The Code emphasizes that the receptacle under the sink must be accessible and located to avoid physical damage to the flexible cord used to connect the motor to the power source (Fig. 12-25).

The cord used for this type of installation should be Type S, SO, ST, STO, SJO, SJT, SJTO, or SPT-3, which is a three-conductor cord terminating with a grounding-type plug. The cord must be at least 18 and not more than 36 inches long.

The hookup of trash compactors and dishwashers is the same as the hookup for garbage disposers. However, the cord must be from 3 to 4 feet long, instead of 1½ to 3 feet, as for the disposers.

The ampere rating of an individual branch circuit to furnish power to these appliances must not be less than the marked ampere rating of the appliance.

Each of the receptacles serving an appliance with a flexible cord must have some way of disconnecting

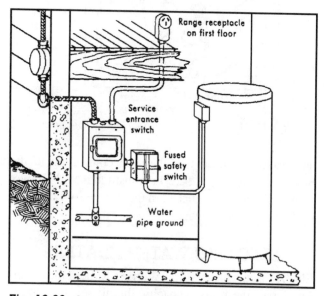

Fig. 12-23 *One method of installation of an electric hot water heater.*

Labels in figure: Range receptacle on first floor; Service entrance switch; Fused safety switch; Water pipe ground

Table 12-3 Wire Sizes for Single-Phase Motor Circuits

Horsepower of Motor	Volts	Approximate Starting Current (Amperes)	Approximate Full Load Current (Amperes)	Length of Run (Feet) from Main Switch to Motor							
				Feet	25	50	75	100	150	200	300	400
$1/4$	120	20	5	Wire size	14	14	14	12	10	10	8	6
$1/3$	120	20	5.5	Wire size	14	14	14	12	10	8	6	6
$1/2$	120	22	7	Wire size	14	14	12	12	10	8	6	6
$3/4$	120	28	9.5	Wire size	14	12	12	10	8	6	4	4
$1/4$	240	10	2.5	Wire size	14	14	14	14	14	14	12	12
$1/3$	240	10	3	Wire size	14	14	14	14	14	14	12	10
$1/2$	240	11	3.5	Wire size	14	14	14	14	14	12	12	10
$3/4$	240	14	4.7	Wire size	14	14	14	14	14	12	10	10
1	240	16	5.5	Wire size	14	14	14	14	14	12	10	10
$1\,1/2$	240	22	7.6	Wire size	14	14	14	14	12	10	8	8
2	240	30	10	Wire size	14	14	14	12	10	10	8	6
3	240	42	14	Wire size	14	12	12	12	10	8	6	6
5	240	69	23	Wire size	10	10	10	8	8	6	4	4
$7\,1/2$	240	100	34	Wire size	8	8	8	8	6	4	2	2
10	240	130	43	Wire size	6	6	6	6	4	4	2	1

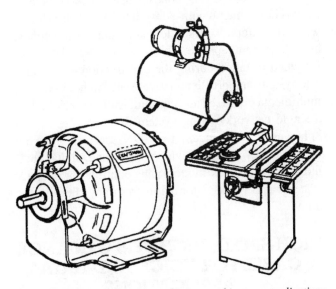

Fig. 12-24 *Electric motors are being used in many applications throughout the home.*

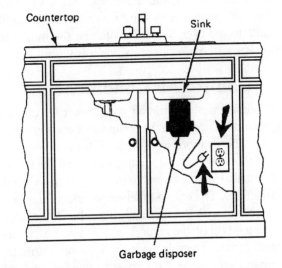

Fig. 12-25 *Installation of garbage disposers under the sink. Remember, double-insulated disposers do not require a ground.*

the circuit so that it may be serviced or repaired by removing the plug from the receptacle. Most appliances need service from time to time. Thus, it is a good idea to make sure the repair person is properly protected by the removal of the plug from the receptacle. It should have an easily accessible receptacle so the plug can be removed before work begins on the appliance.

The Code permits electric ranges to be supplied by cord and plug connections to a range receptacle located at the rear base of the range. The rule allows such a plug and receptacle to serve as the means for disconnecting the range if the connection is accessible from the front by the removal of a drawer.

In some cases, a gas range is used and the clock and oven lights are electric. Then the receptacle for the lights and clock must be easily accessible without having to disassemble the range.

AIR CONDITIONERS

Electrical wiring for a single-unit air conditioner, window-mounted, is treated as an individual single-motor load of ampere rating as marked on the nameplate.

If the Code is checked, it can be found that, according to Section 430-53(a), two 115-volt air conditioners of 6 amperes each can be used on a 15-ampere circuit. Three 6-ampere conditioners can be used on a

20-ampere circuit if the circuit does not supply any other load. Two 220-volt, 6-ampere units could be operated on a 15-ampere circuit.

Most air conditioners today exceed the 6-ampere full-load current rating. Therefore, the application of more than one air conditioner on a 15-ampere line is limited.

Many local codes avoid the complications of connecting conditioners to existing circuits and connecting more than one conditioner to the same circuit by requiring a separate branch circuit for each conditioner.

If a unit room air conditioner has a fixed (not a cord and plug) connection to its supply, it must be treated as a group of several individual motors. It must be protected in accordance with the sections covering several motors on one branch circuit.

A disconnect is required for every unit air conditioner. The plug and receptacle or a separate connector may serve as the disconnect.

If a fixed unit (no plug and receptacle) is wired into a circuit, the switch or circuit breaker must be readily available to the user. The off switch on the individual unit serves as the disconnect, since it is within reach.

Central Air Conditioners

Some homes use a central air conditioner for cooling. This requires a unit with the compressor and condenser outside. The evaporator is usually located inside. A duct system carries the cool air throughout the house.

Electrical wiring in and to the units varies with the manufacturer. The extent to which the electrician must be concerned with the fuse and circuit breaker calculations depends upon the manner in which the unit motors are fed. It also depends upon the type of distribution system to which the unit is to be connected. A packaged unit is treated as a group of motors. This is different from the approach used for the window units.

Some equipment will be delivered with the branch circuit selection current marked. This simplifies the situation. All the required controls and the size of the wire can be judged according to this information.

In sizing the wire for this type of fixed unit, the selection of components for control must be the *rated-load current* marked on the equipment or the compressor.

The disconnect for a hermetic motor (used in the compressor of the air conditioner) must be a motor circuit switch rated in horsepower or a circuit breaker.

If a circuit breaker is used, it must have an ampere rating of not less than 115% of the nameplate rated-load current or the branch circuit selection current. The larger of the two ratings would be taken as a working base for sizing the wire.

NEWER WIRING SYSTEMS

The current method of wiring a house has not changed for decades. What has been and is usually installed is simple quad wiring supporting plain old telephone service and a low-performance coaxial cable for cable TV service. This cabling provides support for up to two basic, analog telephone lines and a limited number of CATV channels. This wiring situation does not allow for any home systems integration and does not include forethought for the future. The characteristics of a home wiring network need to be carefully defined. The products comprising a home network should be specifically designed for residential use. They should be easy to install in a new home construction as well as in existing homes. The system should be modular and flexible to fit the homeowner's needs. A common in-home network provides great flexibility and lower overall cost than this separate dedicated wiring.

One of the Bell Telephone Companies newer organizations has come up with a design for the modern home with computers, VCRs, television, and fax machines, as well as the telephone. This system is especially useful when installed during the construction of a house. The system supports, interactive voice, high-speed data, multimedia products, and communication services. It is a single network wiring system that provides instant "plug and play" access to ISDN services, Internet access, CATV programs, video on demand, digital satellite signals, and fax/modem plus controls for security systems and home automation from anywhere in the home.

HIGH-SPEED, HIGH-PERFORMANCE CABLE FOR VOICE AND DATA APPLICATIONS

Today's evolving technologies are requiring line speeds that are fast—up to 155 Mbps. The uses are diverse—not just voice communication, but also fax, electronic mail, video, data, and file transfer. Applications include everything from telecommuting to video conferencing to home-based businesses. To help the home keep pace with this changing communications mix, a high-quality coaxial cable must be installed in new homes. This high-speed, high-performance cable carries a full range of high-speed communications services up to 100 meters in home distribution systems. See Figs. 12-26 and 12-27. Two colors are used for the cables. Black is used for external video applications such as cable television, and white is used for internal video applications such as security cameras.

Fig. 12-26 *Twisted-pair cable shown after connectorizaton. (Lucent)*

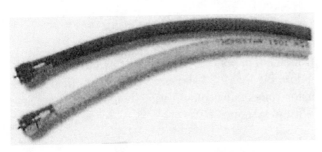
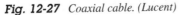

Fig. 12-27 *Coaxial cable. (Lucent)*

Installation of Cable

The primary cable installed between the service center and outlets consists of one four-pair Category 5 cable and two RG6 coaxial cables in a single jacket. See Fig. 12-28. All cable runs are direct from the service center to the outlet. One of the two RG6 coaxial cables in each home run is external cable and carries the output signal that drives TVs and VCRs. It is the electrical combination of the CATV input from the cable TV company and all input signals brought to the service center on the other RG6 internal cable. The RF modulated output of VCRs, CD players, and security cameras is sent to the service center on the internal coaxial cables. Up to 16 internal cable signals are combined at the service center, and then are mixed with the CATV input to form the external signals.

Service Center

The service center (Fig. 12-29) is a centralized distribution point that connects communications devices to a single, uniform structure cabling system in the home. This service center contains distribution devices for voice data, and both baseband and broadband video. Up to 31 twisted pair connections can be made to the service center when it is fully configured. The unshielded twisted pair distribution can service telephone service, LAN transmission, or computer modems. The service center has many video applications: cable television, satellite dishes, or video cameras for in-home security. They can be connected through the mounting panels in the service center. This family of panels holds video splitters and combiners that service up to 16 dual video outlets when fully configured. A distribution amplifier is also located in the service center. It boosts the cable television feed signal and combines it with in-home sources for distribution throughout the house. It is a

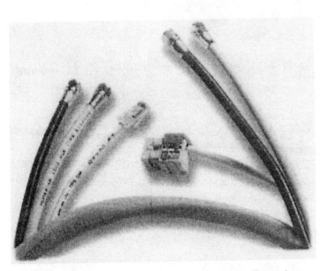

Fig. 12-28 *Hybrid cable for home wiring. (Lucent)*

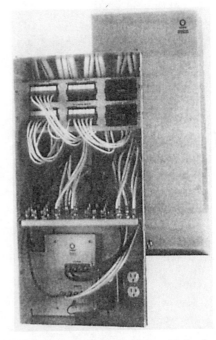

Fig. 12-29 *Service center. (Lucent Technologies)*

necessary component when the home has more than four ports, or has cable runs exceeding 150 feet from the service center.

Figure 12-30 shows a mounting panel while Fig. 12-31 illustrates the distribution amplifier. The amplifier is mounted on a panel, such as that in Fig. 12-31. The amplified video distribution is necessary for excellent picture quality. Figure 12-32 shows the mounting for the service center. It can be surface- or flush-mounted, and the distribution modules are preassembled with front-facing fasteners. The enclosure and all components are grounded for safety. Cable can enter the enclosure

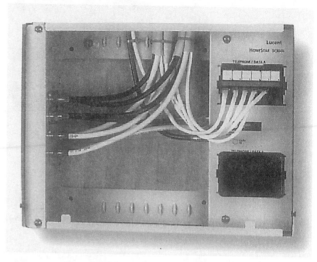

Fig. 12-32 *Service center box. (Lucent)*

Fig. 12-30 *Mounting panel. (Lucent)*

from the top, bottom, or back for flexible, easy installation. Figure 12-33 shows a crimper and field terminator modulator plugs used with the HomeStar Wiring* system.

Fig. 12-31 *Distribution amplifier and AC mounting panel. (Lucent)*

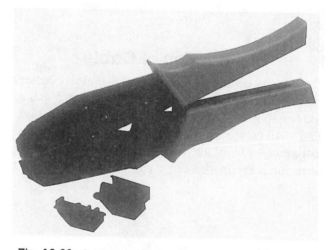

Fig. 12-33 *Crimper and field terminator for modular plugs used in HomeStar wiring. (Lucent)*

*Lucent Technologies registered trademark.

APPENDIX

Tables

Table A-1 *Physiological Effects of Electric Currents**

	Readings	Effects
Safe current values	1 mA or less	Causes no sensation—not felt.
	1 to 8 mA	Sensation of shock, not painful; individual can let go at will since muscular control is not lost.
	8 to 15 mA	Painful shock; individual can let go at will since muscular control is not lost.
	15 to 20 mA	Painful shock; control of adjacent muscles is lost; victim cannot let go.
Unsafe current values	50 to 100 mA	Ventricular fibrillation—a heart condition that can result in instant death—is possible.
	100 to 200 mA	Ventricular fibrillation occurs.
	200 mA or over	Severe burns, severe muscular contractions—so severe that chest muscles clamp the heart and stop it for the duration of the shock. (This prevents ventricular fibrillation.)

*Information provided by National Safety Council.

Table A-2 *SI Prefixes*

Multiplying Factor	Prefix	Symbol
$1,000,000,000,000 = 10^{12}$	tera	T
$1,000,000,000 = 10^{9}$	giga	G
$1,000,000 = 10^{6}$	mega	M
$1,000 = 10^{3}$	kilo	k
$100 = 10^{2}$	hekto	h
$10 = 10^{1}$	deka	da
$0.1 = 10^{-1}$	deci	d
$0.01 = 10^{-2}$	centi	c
$1.001 = 10^{-3}$	milli	m
$0.000.001 = 10^{-6}$	micro	μ
$0.000.000.001 = 10^{-9}$	nano	n
$0.000.000.000.001 = 10^{-12}$	pico	p
$0.000.000.000.000.001 = 10^{-15}$	femto	f
$0.000.000.000.000.000.001 = 10^{-18}$	atto	a

Table A-3 *Powers of 10*

$$1 = 10^{0}$$
$$10 = 10^{1}$$
$$100 = 10^{2}$$
$$1,000 = 10^{3}$$
$$10,000 = 10^{4}$$
$$100,000 = 10^{5}$$
$$1,000,000 = 10^{6}$$

Likewise, powers of 10 can be used to simplify decimal expressions. The submultiples of 10 from 0.1 to 0.000001, with their equivalents in powers of 10, are

$$0.1 = 10^{-1}$$
$$0.01 = 10^{-2}$$
$$0.001 = 10^{-3}$$
$$0.0001 = 10^{-4}$$
$$0.00001 = 10^{-5}$$
$$0.000001 = 10^{-6}$$

Table A-4 *Conversion Factors in Converting from U.S. Customary System Units to Metric Units*

To find	Multiply	By
micrometers	mils	25.4
centimeters	inches	2.54
meters	feet	0.3048
meters	yards	0.9144
kilometers	miles	1.609344
grams	ounces	28.349523
kilograms	pounds	0.4539237
liters	gallons (U.S.)	3.7854118
liters	gallons (Imperial)	4.546090
milliliters (cc)	fluid ounces	29.573530
milliliters (cc)	cubic inches	16.387064
square centimeters	square inches	6.4516
square meters	square feet	0.09290304
square meters	square yards	0.83612736
cubic meters	cubic feet	2.8316847×10^{-2}
cubic meters	cubic yards	0.76455486
joules	Btu	1054.3504
joules	foot-pounds	1.35582
kilowatts	Btu per minute	0.01757251
kilowatts	foot-pounds per minute	2.2597×10^{-5}
kilowatts	horsepower	0.7457
radians	degrees	0.017453293
watts	Btu per minute	17.5725

Table A-5 *Metric and U.S. Customary System Equivalents*

Measures of Length

1 meter = $\begin{cases} 39.3 \text{ inches} \\ 3.28083 \text{ feet} \\ 1.0936 \text{ yards} \end{cases}$

1 centimeter = 0.3937 inch

1 millimeter = 0.03937 inch, or nearly 1/25 inch

1 kilometer = 0.62137 mile

1 foot = 0.3048

1 inch = $\begin{cases} 2.54 \text{ centimeters} \\ 25.40 \text{ millimeters} \end{cases}$

Measures of Surface

1 square meter = 10.764 square feet
= 1.196 square yards

1 square centimeter = 0.155 square inch

1 square millimeter = 0.00155 square inch

1 square yard = 0.836 square meter

1 square foot = 0.0929 square meter

1 square inch = $\begin{cases} 6.452 \text{ square centimeters} \\ 645.2 \text{ square millimeters} \end{cases}$

Measures of Volume and Capacity

1 cubic meter = $\begin{cases} 35.314 \text{ cubic feet} \\ 1.308 \text{ cubic yards} \\ 264.2 \text{ U.S. gallons (231 cubic inches)} \end{cases}$

1 cubic decimeter = $\begin{cases} 61.0230 \text{ cubic inches} \\ 0.0353 \text{ cubic feet} \end{cases}$

1 cubic centimeter = 0.061 cubic inch

1 liter = $\begin{cases} 1 \text{ cubic decimeter} \\ 61.0230 \text{ cubic inches} \\ 0.0353 \text{ cubic foot} \\ 1.0567 \text{ quarts (U.S.)} \\ 0.2642 \text{ gallon (U.S.)} \\ 2.2020 \text{ pounds of water at } 62°F \end{cases}$

1 cubic yard = 0.7645 cubic meter

1 cubic foot = $\begin{cases} 0.02832 \text{ cubic meter} \\ 28.317 \text{ cubic decimeters} \\ 28.317 \text{ liters} \end{cases}$

1 cubic inch = 16.383 cubic centimeters

1 gallon (British) = 4.543 liters

1 gallon (U.S.) = 3.785 liters

Table A-6 Temperature Conversion

The numbers in italics in the center column refer to the temperature, in either degrees Celsius or degrees Fahrenheit, that is to be converted to the other scale. If you are converting a Fahrenheit temperature to Celsius, find the number in the center column and then look to the left column for the Celsius temperature. If converting a Celsius temperature, find the number in the center column and look to the right for the Fahrenheit equivalent.

−100 to 30			31 to 71			72 to 212			213 to 620			621 to 1000		
°C		°F	°C		°F	°C		°F	°C		°F °	C		°F
−73	−100	−148	−0.6	31	87.8	22.2	72	161.6	104	220	428	332	630	1166
−68	−90	−130	0	32	89.6	22.8	73	163.4	110	230	446	338	640	1184
−62	−80	−112	0.6	33	91.4	23.3	74	165.2	116	240	464	343	650	1202
−57	−70	−94	1.1	34	93.2	23.9	75	167.0	121	250	482	349	660	1220
−51	−60	−76	1.7	35	95.0	24.4	76	168.8	127	260	500	354	670	1238
−46	−50	−58	2.2	36	96.8	25.0	77	170.6	132	270	518	360	680	1256
−40	−40	−40	2.8	37	98.6	25.6	78	172.4	138	280	536	366	690	1274
−34.4	−30	−22	3.3	38	100.4	26.1	79	174.2	143	290	554	371	700	1292
−28.9	−20	−4	3.9	39	102.2	26.7	80	176.0	149	300	572	377	710	1310
−23.3	−10	14	4.4	40	104.0	27.2	81	177.8	154	310	590	382	720	1328
−17.8	0	32	5.0	41	105.8	27.8	82	179.6	160	320	608	388	730	1346
−17.2	1	33.8	5.6	42	107.6	28.3	83	181.4	166	330	626	393	740	1364
−16.7	2	35.6	6.1	43	109.4	28.9	84	183.2	171	340	644	399	750	1382
−16.1	3	37.4	6.7	44	111.2	29.4	85	185.0	177	350	662	404	760	1400
−15.6	4	39.2	7.2	45	113.0	30.0	86	186.8	182	360	680	410	770	1418
−15.0	5	41.0	7.8	46	114.8	30.6	87	188.6	188	370	698	416	780	1436
−14.4	6	42.8	8.3	47	116.6	31.1	88	190.4	193	380	716	421	790	1454
−13.9	7	44.6	8.9	48	118.4	31.7	89	192.2	199	390	734	427	800	1472
−13.3	8	46.4	9.4	49	120.0	32.2	90	194.0	204	400	752	432	810	1490
−12.8	9	48.2	10.0	50	122.0	32.8	91	195.8	210	410	770	438	820	1508
−12.2	10	50.0	10.6	51	123.8	33.3	92	197.6	216	420	788	443	830	1526
−11.7	11	51.8	11.1	52	125.6	33.9	93	199.4	221	430	806	449	840	1544
−11.1	12	53.6	11.7	53	127.4	34.4	94	201.2	227	440	824	454	850	1562
−10.6	13	55.4	12.2	54	129.2	35.0	95	203.0	232	450	842	460	860	1580
−10.0	14	57.2	12.8	55	131.0	35.6	96	204.8	238	460	860	466	870	1598
−9.4	15	59.0	13.3	56	132.8	36.1	97	206.6	243	470	878	471	880	1616
−8.9	16	60.8	13.9	57	134.6	36.7	98	208.4	249	480	896	477	890	1634
−8.3	17	62.6	14.4	58	136.4	37.2	99	210.2	254	490	914	482	900	1652
−7.8	18	64.4	15.0	59	138.2	37.8	100	212.0	260	500	932	488	910	1670
−7.2	19	66.2	15.6	60	140.0	43	110	230	266	510	950	493	920	1688
−6.7	20	68.0	16.1	61	141.8	49	120	248	271	520	968	499	930	1706
−6.1	21	69.8	16.7	62	143.6	54	130	266	277	530	986	504	940	1724
−5.6	22	71.6	17.2	63	145.4	60	140	284	282	540	1004	510	950	1742
−5.0	23	73.4	17.8	64	147.2	66	150	302	288	550	1022	516	960	1760
−4.4	24	75.2	18.3	65	149.0	71	160	320	293	560	1040	521	970	1778
−3.9	25	77.0	18.9	66	150.8	77	170	338	299	570	1058	527	980	1796
−3.3	26	78.8	19.4	67	152.6	82	180	356	304	580	1076	532	990	1814
−2.8	27	80.6	20.0	68	154.4	88	190	374	310	590	1094	538	1000	1832
−2.2	28	82.4	20.6	69	156.2	93	200	392	316	600	1112			
−1.7	29	84.2	21.1	70	158.0	99	210	410	321	610	1130			
−1.1	30	86.0	21.7	71	159.8	100	212	414	327	620	1148			

Table A-7 *Standards for Wire Gages (Dimensions of Sizes in Decimal Parts of an Inch)*

No. of Wire	American or Brown & Sharpe for Nonferrous Metals	Birmingham or Stub's Iron Wire	American S. & W. Co.'s (Washburn & Moen) Std Steel Wire	American S. & W. Co.'s Music Wire	Imperial Wire	Stub's Steel Wire	U. S. Std. Gage for Sheet & Plate Iron & Steel	No. of Wire
7–0	0.651354	...	0.4900	...	0.500	...	0.500	7–0
6–0	0.580049	...	0.4615	0.004	0.464	...	0.46875	6–0
5–0	0.516549	0.500	0.4305	0.005	0.432	...	0.4375	5–0
4–0	0.460	0.454	0.3938	0.006	0.400	...	0.40625	4–0
000	0.40964	0.425	0.3625	0.007	0.372	...	0.375	000
00	0.3648	0.380	0.3310	0.008	0.348	...	0.34375	00
0	0.32486	0.340	0.3065	0.009	0.324	...	0.3125	0
1	0.2893	0.300	0.2830	0.010	0.300	0.227	0.28125	1
2	0.25763	0.284	0.2625	0.011	0.276	0.219	0.265625	2
3	0.22942	0.259	0.2437	0.012	0.252	0.212	0.250	3
4	0.20431	0.238	0.2253	0.013	0.232	0.207	0.2234375	4
5	0.18194	0.220	0.2070	0.014	0.212	0.204	0.21875	5
6	0.16202	0.203	0.1920	0.016	0.192	0.201	0.203125	6
7	0.14428	0.180	0.1770	0.018	0.176	0.199	0.1875	7
8	0.12849	0.165	0.1620	0.020	0.160	0.197	0.171875	8
9	0.11443	0.148	0.1483	0.022	0.144	0.194	0.15625	9
10	0.10189	0.134	0.1350	0.024	0.128	0.191	0.140625	10
11	0.090742	0.120	0.1205	0.026	0.116	0.188	0.125	11
12	0.080808	0.109	0.1055	0.029	0.104	0.185	0.109375	12
13	0.07196	0.095	0.0915	0.031	0.092	0.182	0.09375	13
14	0.064084	0.083	0.0800	0.033	0.080	0.180	0.078125	14

Table A-8 Allowable Ampacity for Flexible Cords and Cables [Based on Ambient Temperature of 30°C (86°F).]

Size (AWG)	Thermoplastic Types TPT, TST	Thermoset Types C, E, EO, PD, S, SJ, SJO, SJOW, SJOO, SJOOW, SO, SOW, SOO, SOOW, SP-1, SP-2, SP-3, SRD, SV, SVO, SVOO / Thermoplastic Types ET, ETLB, ETP, ETT, SE, SEW, SEO, SEOW, SEOOW, SJE, SJEW, SJEO, SJEOW, SJEOOW, SJT, SJTW, SJTO, SJTOW, SJTOO, SJTOOW, SPE-1, SPE-2, SPE-3, SPT-1, SPT-1W, SPT-2, SPT-2W, SPT-3, ST, SRDE, SRDT, STO, STOW, STOO, STOOW, SVE, SVEO, SVT, SVTO, SVTOO		Types HPD, HPN, HSJ, HSJO, HSJOO
		A+	B+	
27*	0.5	—	—	—
20	—	5**	***	—
18	—	7	10	10
17	—	—	12	13
16	—	10	13	15
15	—	—	—	17
14	—	15	18	20
12	—	20	25	30
10	—	25	30	35
8	—	35	40	—
6	—	45	55	—
4	—	60	70	—
2	—	80	95	—

*Tinsel cord.

**Elevator cables only.

***7 amperes for elevator cables only; 2 amperes for other types.

+The allowable currents under subheading A apply to 3-conductor cords and other multiconductor cords connected to utilization equipment so that only 3 conductors are current-carrying. The allowable currents under subheading B apply to 2-conductor cords and other multiconductor cords connected to utilization equipment so that only 2 conductors are current carrying.

Table A-9 Size of Extension Cords for Portable Electric Tools

For 120-Volt Tools						
Full-Load Ampere Rating of Tool	0 to 2.0 A	2.1 to 3.4 A	3.5 to 5 A	5.1 to 7 A	7.1 to 12 A	12.1 to 16 A
Length of Cord	Wire Size (AWG)					
25 feet	18	18	18	16	14	14
50 feet	18	18	18	16	14	12
75 feet	18	18	16	14	12	10
100 feet	18	16	14	12	10	8
200 feet	16	14	12	10	8	6
300 feet	14	12	10	8	6	4
400 feet	12	10	8	6	4	4
500 feet	12	10	8	4	6	2
600 feet	10	8	6	4	2	2
800 feet	10	8	6	4	2	1
1000 feet	8	6	4	2	1	0

Note: If the voltage is already low at the source (outlet), increase to standard voltage or use a much larger cable than listed in order to prevent any further loss in voltage.

Reprinted with permission from NFPA, National Electrical Code, Copyright © 2002, National Fire Protection Association, Quincy, MA 02269. This reprinted material is not the complete and official position of the NFPA on the referenced subject which is represented only by the standard in its entirety.

Table A-10 Demand Factors for Household Electric Clothes Dryers

Number of Dryers	Demand Factor (Percent)
1–4	100%
5	85%
6	75%
7	65%
8	60%
9	55%
10	50%
11	47%
12–22	% = 47 − (number of dryers − 11)
23	35%
24–42	% = 35 − [0.5 × number of dryers − 23)]
43 and over	25%

Table A-11 *Demand Loads for Household Electric Ranges, Wall-Mounted Ovens, Counter-Mounted Cooking Units, and Other Household Cooking Appliances over 1³/₄-kW Rating. (Column A to be used in all cases except as otherwise permitted in Note 3 below.)*

| Number of Appliances | Demand Factor (Percent) (See Notes) | | Column C Maximum Demand (kW) (See Notes) (Not over 12 kW Rating) |
	Column A (Less than 3½ kW Rating)	Column B (3½ kW to 8¾ kW Rating)	
1	80	80	8
2	75	65	11
3	70	55	14
4	66	50	17
5	62	45	20
6	59	43	21
7	56	40	23
8	53	36	23
9	51	35	24
10	49	34	25
11	47	32	26
12	45	32	27
13	43	32	28
14	41	32	29
15	40	32	30
16	39	28	31
17	38	28	32
18	37	28	33
19	36	28	34
20	25	28	35
21	34	26	36
22	33	26	37
23	32	26	38
24	31	26	39
25	30	26	40
26–30	30	24	15 kW + 1 kW for each range
31–40	30	22	
41–50	30	20	25 kW + ¾ kW for each range
51–60	30	18	
61 and over	30	16	

1. Over 12 kW through 27 kW ranges all of same rating. For ranges individually rated more than 12 kW but not more than 27 kW, the maximum demand in Column C shall be increased 5 percent for each additional kilowatt of rating or major fraction thereof by which the rating of individual ranges exceeds 12 kW.

2. Over 8¾ kW through 27 kW ranges of unequal ratings. For ranges individually rated more than 8¾ kW and of different ratings, but none exceeding 27 kW, an average value of rating shall be computed by adding together the ratings of all ranges to obtain the total connected load (using 12 kW for any range rated less than 12 kW) and dividing by the total number of ranges. Then the maximum demand in Column C shall be increased 5 percent for each kilowatt or major fraction thereof by which this average value exceeds 12 kW.

3. Over 1¾ kW through 8¾ kW. In lieu of the method provided in Column C, it shall be permissible to add the nameplate ratings of all household cooking appliances rated more than 1¾ kW but not more than 8¾ kW and multiply the sum by the demand factors specified in Column A or B for the given number of appliances. Where the rating of cooking appliances falls under both Column A and Column B, the demand factors for each column shall be applied to the appliances for that column, and the results added together.

4. Branch-Circuit Load. It shall be permissible to compute the branch-circuit load for one range in accordance with Table 220.19. The branch-circuit load for one wall-mounted oven or one counter-mounted cooking unit shall be the nameplate rating of the appliance. The branch-circuit load for a counter-mounted cooking unit and not more than two wall-mounted ovens, all supplied from a single branch circuit and located in the same room, shall be computed by adding the nameplate rating of the individual appliances and treating this total as equivalent to one range.

5. This table also applies to household cooking appliances rated over 1¾ kW and used in instructional programs.

Table A-12 *Formulas for Motor Applications*

T = torque or twisting moment (force × moment arm length)
π = 3.1416
N = revolutions per minute
HP = horsepower (33,000 ft · lb/min) applies to power output
R = radius of pulley, feet
E = input voltage
I = current, amperes
P = power input, watts

$$HP = \frac{T\,(\text{lb.}\cdot\text{in.}) \times N\,(\text{rpm})}{63{,}025}$$

$$HP = T\,(\text{oz}\cdot\text{in.}) \times N \times 9.917 \times 10^{-7}$$
$$= \text{approx. } T\,(\text{oz}\cdot\text{in.}) \times N \times 10^{-6}$$

$$P = EI \times \text{power factor} = \frac{HP \times 746}{\text{motor efficiency}}$$

Power to Drive Pumps

$$HP = \frac{\text{gal. per min.} \times \text{total head (including friction)}}{3960 \times \text{eff. of pump}}$$

where

$$\text{Approx. friction head (ft)} = \frac{\text{pipe length (ft)} \times [\text{velocity of flow (fps)}]^2 \times 0.02}{5.367 \times \text{diameter (in.)}}$$

Eff. = Approx. 0.50 to 0.85

Time to Change Speed of Rotating Mass

$$\text{Time (s)} = \frac{WR^2 \times \text{change in rpm}}{308 \times \text{torque (ft·lb)}}$$

where

$$WR^2\,(\text{disk}) = \frac{\text{weight (lb)} \times [\text{radius (ft)}^2]}{2}$$

$$WR^2\,(\text{rim}) = \frac{\text{wt. (lb)} \times \{[\text{outer radius (ft)}]^2 + [\text{inner radius (ft)}]^2\}}{2}$$

Power to Drive Fans

$$HP = \frac{\text{air (ft}^3/\text{min)} \times \text{water gage pressure (in.)}}{6.350 \times \text{Eff.}}$$

Index

ABOUT THE AUTHORS

Rex Miller is Professor Emeritus of Industrial Technology at State University College at Buffalo and has taught technical curriculum at the college level for more than 40 years. He is the coauthor of the best-selling *Carpentry & Construction,* now in its fourth edition, and the author of more than 75 texts for vocational and industrial arts programs. He lives in Round Rock, Texas.

Mark R. Miller is Chairman and Associate Professor of Industrial Technology at Texas A&M University in Kingsville, Texas. He teaches construction courses for future middle managers in the trade. He is coauthor of several technical books, including the best-selling *Carpentry & Construction,* now in its fourth edition. He lives in Kingsville, Texas.

Glenn E. Baker is Professor Emeritus of Industrial Technology at Texas A&M University in College Station, Texas. He is the author of a number of books, including *Carpentry & Construction,* Fourth Edition. He lives in College Station, Texas.